INTERNATIONAL EXHIBITION, 1876.

BOARD ON BEHALF OF UNITED STATES EXECUTIVE DEPARTMENTS.

CLASSIFICATION

OF THE

COLLECTION TO ILLUSTRATE

THE

ANIMAL RESOURCES OF THE UNITED STATES.

A LIST OF SUBSTANCES DERIVED FROM THE ANIMAL KINGDOM, WITH SYNOPSIS OF THE USEFUL AND INJURIOUS ANIMALS AND A CLASSIFICATION OF THE METHODS OF CAPTURE AND UTILIZATION.

By G. BROWN GOODE, M. A

ASSISTANT CURATOR U. S. NATIONAL MUSEUM.

WASHINGTON:
GOVERNMENT PRINTING OFFICE.
1876.

ADVERTISEMENT.

This work is the sixth of a series of papers intended to illustrate the collections of Natural History and Ethnology belonging to the United States and constituting the National Museum, of which the Smithsonian Institution was placed in charge by the act of Congress of August 10, 1846.

It has been prepared at the request of the Institution, and printed by authority of the honorable Secretary of the Interior.

JOSEPH HENRY,
Secretary Smithsonian Institution.

SMITHSONIAN INSTITUTION,
Washington, February, 1876.

PREFACE.

The following classification has been prepared by Mr. Goode to facilitate the work of collecting and arranging the material gathered by the National Museum to illustrate the resources of the United States as derived from the animal kingdom, in the International Exhibition of 1876. It is also intended to indicate the general character of the articles which are to be included in this branch of the exhibition.

Contributions of specimens of the different classes enumerated are much desired, for the purpose of making the proposed display complete, and should be addressed to the Smithsonian Institution, Washington, D. C.

JOSEPH HENRY,

TABLE OF CONTENTS.

*** Centrifugal missiles, (propelling power augmented by artificial lengthening of the arm.)*

SECTION D.—ANIMAL PRODUCTS AND THEIR APPLICATIONS.

INTRODUCTION.

The system proposed in the following lists has been hurriedly prepared, and is necessarily very incomplete; it is intended merely as a provisional classification, to be used in collecting the materials for the exhibition, and in their preliminary arrangement.

The first group, SECTION A, is an index to the whole series; it will include all North American animals which are directly beneficial or injurious to man. Although every species, down to the very least, exercises some influence upon human well-being, it seems scarcely practicable to attempt the exhibition of those which affect it only indirectly. Those species are considered useful which supply food, clothing, shelter, implements, materials, and amusement; those injurious which endanger the life or personal comfort of man, or destroy those animals and plants which are of direct benefit to him. In the enumeration of animals, the names of the orders are given, followed in parenthesis by the best-known names of the more important species included, and a brief note on their principal uses. This enumeration, being simply of convenience, makes no claim to zoological precision.

SECTION B embraces all instruments and methods employed by the hunters, trappers, and fishermen of North America, aboriginal and civilized. Not only those which are directly employed in destruction or capture are included, but the means made use of in pursuing or attracting the animals and fishes, and the personal equipment of the pursuer. The collection will be a monograph of *all* matters relating to the chase and the fisheries of the country. In preparing the classification here submitted, the principles of zoological classification have been followed as closely as possible; each distinct form has been considered a species; and the specific forms have been grouped into genera, families, and orders according to the general balance of their affinities. Form and manner of use have not been without weight, but superficial resemblance has been set aside, and the *idea* given the first importance. Thus, barbed spears and harpoons have been placed with the "hooked instruments," while plain spears and lances are grouped with the knives and clubs. In studying the place of the fire-arms and bows and arrows, the missile itself has been regarded as more important than the machine which propels it, and the latter is placed in a subordinate relation.

In one group, that of nets, convenience in arrangement of the specimens seems to demand that *material*, a character of small importance, shall be made prominent. Two widely diverging groups of apparatus are associated under the head of nets, viz, encircling-nets, the true relations of which are with grasping and scooping instruments, and entangling nets which belong with the traps, where a third group of nets, the pound and weir nets are actually classed. Where the exigencies of administration of the specimens demand that they should be arranged otherwise than in their exact systematic position, full cross-references are given.

The simplest implements have always been placed first, the series advancing in the order of complication of structure. Thus we have in the beginning the apparatus of direct application, or tools, including, first, those implements which are used in the hand, and which increase its power in a simple way, such as clubs and slung-shot, which merely add to the weight of the fist, followed by the knives, axes, and spears, which in their simplest and primitive form were sharpened stones and pointed sticks. Second, are the grasping-implements, or those by which the power of the fingers is extended. In this series the same principle of progress from simple to complex is followed; in the scoop we have the idea of the hollow palm of the hand developed in various forms, while the grasping-hooks and grasping-lines are the artificial extensions of the human finger. Under hooked instruments, the simple hooks, or those which are attached to the object by a single motion, a pulling one, are placed first, followed by the barbed implements, in which the attachment is made by a thrusting, succeeded by a pulling motion, and then by the tongs and forceps, which are essentially double hooks. The succeeding division is that containing the lasso and bolas, which are worked at long distances and require great skill, succeeded by the tangles, which are, in principle, assemblages of lassos, entangling objects among their fiber nooses.

In the third division, that of missiles, the same principle of succession is adhered to. First are placed those missiles which are propelled by the unaided arm; then those in using which the arm is artificially lengthened, as with the sling, string, or darting stick; then those in which the propelling power is derived from the elasticity of rods and cords, the strength of the arm having become subsidiary; closing with those in the use of which the strength of the arm is of no essential value, and the propelling power originates in chemical combustion.

Accessory to these are groups containing those articles used in the manufacture, testing, loading, and transportation of these missiles, and the machines which drive them through the air.

In a fourth division is the apparatus of angling, which is separated from hooked implements with which the form of the articles would naturally place them, since they are not implements of grasping, but partake of the nature of traps, being in part automatic.

The group of nets is a heterogeneous one, as has been stated above, consisting of two divisions, the first that of entangling-nets, belonging properly with traps, while encircling-nets are in idea instruments for grasping.

In arranging traps a logical succession has been preserved as far as possible. Those traps are considered the most simple in which the animal is penned by its own act, without any change in the arrangement of the trap. The pit-falls or "tipes" are first, followed by the mazes or labyrinths of greater or less complexity. Then come the traps in which the entrance is closed, either by the falling of a door or by the falling of a box-like trap, as a whole, so as to surround the animal. Under clutching-traps are placed those which seize the animal, as in the fingers, while crushing-traps are those which seize or impale it bodily. Adhesive preparations, such as bird-lime, close the series.

The accessory divisions, including hunting-animals, decoys, and disguises, and the methods and appliances of pursuit, do not admit any thorough classification, and are arranged with reference to convenience of exhibition.

SECTION C includes all methods of utilizing animals' products. It might be more satisfactorily arranged with the following section, were it not for the inconvenience of exhibiting models and tools in the same cases with the manufactured products; the arrangement of the two sections is nearly the same.

In SECTION D are grouped all useful substances derived from the animal kingdom. In order to avoid the omission of any products which are or may be obtained from North American animals, this enumeration has been made general, those not American being included in parentheses. This enumeration is far from complete, and is intended simply as an aid to future study in the same direction.

SECTION E includes all articles illustrating the culture and protection of useful animals.

SECTION A.

ENUMERATION

OF

ANIMALS BENEFICIAL OR INJURIOUS TO MAN,

(WITH A SYNOPSIS OF THEIR USEFUL APPLICATIONS.)

I. MAMMALS.

1. FERAE:

FISSIPEDIA. (Cats, pumas, jaguars, ocelots, lynxes, wolves and dogs, foxes, fishers, martens, minks, weasels, wolverenes, badgers, skunks, otters, sea-otters, bears, raccoons, and the domesticated cat, dog, and ferret.)

Useful products:

Food, (bears, raccoon, &c.) D. 1.
Fur, (all the group.) D. 6.
Leather, (dog, cat.) D. 20.
Textile fabrics, felt, (raccoon.) D. 8.
Ivory, teeth, (bear, fox, &c.) D. 9.
Claws used by Indians, (bears, puma.) D. 11.
Hair, for brushes, (badger, dog, weasel, skunk, bear.) D. 21.
Oil, (bears.) D. 27.
Perfumes, (civet, &c.) D. 28.
Medicinal products, (skunk.) D. 30.
Chemical agent, *album græcum*, (dog.) D. 30.

Useful traits:

Susceptible of domestication, (wolves (Indian dog,) foxes, otters, bears, raccoon, dog, cat, ferret.)
Employed in hunting, (dog, cat, ferret.) B. 40.
Employed in fishing, (otter.) B. 40.

1. FERAE—Continued.

FISSIPEDIA—Continued.

Injurious traits:

- Enemies of man, (cats, wolves, bears.)
- Enemies of domestic animals.
- Marauders on crops, (bears, raccoon.)
- Stench nuisances, (skunks.)
- Modes of capture. B. I, II, III, VI, X.

PINNIPEDIA. (Fur-seals, sea-lions, hair-seals, hood-seals, sea-elephants, walruses.)

Useful products:

- Food of aborigines. D. 1.
- Fur, (fur-seals, &c.) D. 6.
- Leather, &c., parchment from viscera, (sea-lions, hair-seals, walruses.) D. 20.
- Oil, (hair-seal, hood-seal, sea-elephant, &c.) D. 27.
- Ivory, (walrus.) D. 9.

Injurious traits: Destroy fish.

Modes of capture. B. I, III, V, X.

2. UNGULATA. (Bison, musk-ox, mountain-goat, mountain-sheep, antelope, moose, caribou, elk, deer, peccary, and the domesticated ox, goat, sheep, hog, horse, ass, and camel.)

Useful products:

- Food, fresh, smoked, and pickled, (all the group.) D. 1, 2, 3.
- Fur, (bison, musk-ox, goat, sheep, moose, &c.) D. 6.
- Leather, (all the group.) D. 20.
- Textile fabrics and felt, (ox, goat, sheep, camel, musk-ox.) D. 8.
- Ivory and bone. D. 9.
- Horn, (bison, ox, goat, sheep, deer, elk, &c.) D. 10.
- Hoof, (bison, musk-ox, goat, sheep, deer, horse, &c.) D. 11.
- Hair, bristles, and wool, (bison, ox, goat, sheep, deer, hog, camel.) D. 21.
- Gelatine and glue. D. 4, 24.
- Oil and fat. D. 27.
- Perfumes, (musk-ox, musk-deer.) D. 28.
- Coloring materials from blood and bile. D. 29.

2. UNGULATA—Continued.

Useful products:

Chemical products. D. 30.

Fertilizers. D. 31.

Useful traits: Susceptible of domestication.

Modes of capture. B. I, II, III, VI, IX.

3. PROBOSCIDEA. (Elephants.*)

Useful products: Ivory. D. 9.

4. SIRENIA. (Manatee, or sea-cow.)

Useful products:

Food. D. 1.

Leather. D. 7, 20.

Oil. D. 27.

5. CETE. (Whales.)

DENTICETE. (Beluga, narwhal, porpoise, black-fish, killer, grampus, sperm-whale.)

Useful products:

Food, Indian, (sperm-whale, porpoise.) D. 1.

Oils, (all the group.) D. 27.

Spermaceti, (sperm-whale,)

Leather, (porpoise, beluga.) D. 7, 20.

Bone and ivory, (narwhal, sperm-whale.) D. 9.

Perfume, *ambergris*, (sperm-whale.)

Injurious traits: Destroy fish and seals.

Modes of capture. B. I, II, III, X.

MYSTICETE. (Right, or whale-bone whales.)

Useful products:

Food, (right-whale.) D. 1.

Baleen. D. 12.

Oil, (right-whale, &c.) D. 27.

6. CHIROPTERA. (Bats.)

Useful products:

Food, Indians. D. 1.

Felting material. D. 8.

Guano. D. 31.

Useful traits: Destroy troublesome insects.

Injurious traits: Disseminate troublesome insects.

---

* *Elephas primigenius*, found fossil in North America.

7. INSECTIVORA. (Moles and shrews.)

Useful products:

Fur, (moles.) D. 6.

Felting material, (moles.) D. 8.

Useful traits:

Destroy burrowing insects, &c.

Injurious traits: Burrowers.

Modes of capture. B. VI.

8. GLIRES. (Squirrels, prairie-dogs, showtl, marmots, musquash, beaver, rats, mice, lemmings, porcupines, rabbits, and the domesticated rabbit, and Guinea-pig.)

Useful products:

Food. D. 1.

Fur, (squirrels, showtl, marmots, musquash, beaver, lemmings, rabbit, &c.) D. 6.

Textile fabric, felt, (musquash, beaver, rabbit.) D. 8.

Ivory, (beaver.) D. 9.

Leather, (rat, beaver.) D. 7, 20.

Hair and down, (rabbits.) D. 21.

Quills, (porcupine.) D. 21.

Perfume, *castoreum*, (beaver.) D. 28.

Useful traits: Susceptible of domestication, (squirrels, rabbits, &c.)

Injurious traits: Marauders.

Modes of capture. B. I, II, III, VI, VII, VIII, IX.

9. BRUTA. (Armadillo, &c.)

Useful products: Shell used by Indians in various manufactures. D. 14.

Injurious traits:

Burrower.

Marauder.

10. MARSUPIALIA. (Opossum.)

Useful products:

Food. D. 1.

Hair used in felting. D. 8.

Injurious traits: Marauder.

Modes of capture. B. I, II, III, VI, VIII

II. BIRDS.

11. PASSERES. (Thrushes, stone-chats and blue-birds, dippers, kinglets, titmice, nuthatches, creepers, wrens, larks, wagtails, warblers, tanagers, swallows, waxwings, greenlets, shrikes, finches, starlings, black-birds and orioles, crows and jays, fly-catchers, and domesticated sparrow, canary, &c.)

Useful products:

Foods, (thrushes, rice-birds, &c.) D. 1.

Ornamental feathers. D. 23.

Useful traits:

Destroy insects.

Song-birds, (generally susceptible of domestication.)

12–13. PICARIÆ AND CUCULI. (Night-hawks, whippoorwills, swifts, humming-birds, trogons, saw-bills, kingfishers, cuckoos, woodpeckers.)

Useful products: Ornamental feathers, (humming-birds, trogons.) D. 23.

Useful traits:

Destroy noxious insects, (night-hawks, swifts.)

Destroy tree-borers, (woodpeckers.)

Injurious traits:

Destroy fish, (kingfishers.)

Destroy birds'-eggs, (cuckoos.)

Destroy fruit, (woodpeckers.)

Destroy trees, (sap-sucker.)

14. PSITTACI. (Parroquet and domesticated parrots.)

Useful products: Ornamental feathers. D. 23.

Useful traits: Susceptible of domestication.

15. RAPTORES. (Owls, hawks, eagles, vultures, buzzards.)

Useful products:

Ornamental feathers. D. 23.

Quills. D. 22.

Useful traits:

Susceptible of domestication and use in hunting.[1]

---

[1] Nine species of falcons, hawks, and owls have been employed in the chase by Europeans.

15. RAPTORES—Continued.

Useful traits:

Scavengers, (vultures, buzzards.)

Destroy vermin, (owls, hawks.)

Injurious traits: Destroy domestic animals, eggs, &c.

16. COLUMBÆ. (Pigeons and doves.)

Useful products:

Food.

Ornamental feathers. D. 23.

Useful traits:

Game-birds.

Susceptible of domestication.

Used as targets, (wild pigeon.) B. 25.

Used as carriers, (carrier-pigeon.)

17. GALLINÆ. (Turkey, grouse, partridge, sage-cock, ptarmigan quail, and the domesticated peacock, guinea-fowl, and fowl.)

Useful products:

Foods, flesh. D. 1.

Ornamental feathers. D. 23.

Quills. D. 22.

Albumen. D. 30.

Useful traits:

Game-birds.

Susceptible of domestication.

18. LIMICOLÆ. (Plover, ring-neck, surf-bird, oyster-catcher, turnstone avoset, stilt, phalarope, woodcock, snipe, sandpiper dunlin, godwit, sanderling, willet, tattler, yellow shanks, green-shanks, curlew.)

Useful products:

Food: Flesh. D. 1.

Eggs. D. 1.

Feathers. D. 23.

Useful traits: Game-birds.

19. HERODIONES. (Ibises, spoonbills, herons, egrets, bitterns.)

Useful products: Ornamental feathers. D. 23.

Useful traits: Destroy vermin.

20. ALECTORIDES. (Cranes, rails, crakes, gallinules, coots.)
Useful products:
Food, (rails, crakes.) D. 1.
Feathers. D. 23.
Useful traits: Susceptible of domestication, (cranes.)

21. LAMELLIROSTRES. (Flamingoes, swans, geese, ducks.)
Useful products:
Food: Flesh, (geese, ducks.) D. 1.
Eggs, (geese, ducks.)
Ornamental feathers, (flamingo, geese, &c.) D. 23.
Down, (geese, ducks.) D. 23.
Useful traits:
Susceptible of domestication, (geese, ducks.)
Used as decoys for other swimmers, (brants, ducks.)

22. STEGANOPODES. (Gannets, pelicans, cormorants, darters or water-turkeys, frigate birds, tropic birds.)
Useful products:
Ornamental feathers, (darters, tropic birds.) D. 23.
Leather, (of feet.) D. 20.
Useful traits: Susceptible of domestication.[1]

23. LONGIPENNES. (Gulls, terns, skimmers, petrels, albatrosses shearwaters.)
Useful products:
Food, eggs. D. 1.
Ornamental feathers, (gulls, terns, &c.) D. 23.
Oil, (petrels, &c., used by Eskimos.) D. 27.

24. PYGOPODES. (Loons, grebes, auks, puffins, guillemot, murres.)
Useful products:
Foods, (eggs.) D. 1.
Ornamental feathers, (grebes.) D. 23.
Feathers used as furs, (grebes, auks, &c.) D. 8.

25. SPHENISCI. (Penguins.)
Useful products:
Feathers used as fur. D. 6.
Oil. D. 27.

---

[1] *Graculus carbo* used in Europe for fishing and a similar species in China.

III. REPTILES.

26. CROCODILIA. (Alligator, crocodile.)
    - Useful products:
        - Food. D. 1.
        - Ivory. D. 8.
        - Leather. D. 20.
        - Oil. D. 27.
        - Musk. D. 28.
    - Injurious traits:
        - Enemies of man and domestic animals.
27. TESTUDINATA. (Tortoises, terrapin, leather-back, green, loggerhead, and hawks-bill turtles.)
    - Useful products:
        - Food: Flesh, (green turtle, terrapin, gopher tortoise.) D. 1.
          Eggs, (green turtle, terrapin, gopher tortoise.)
        - Oil from eggs, (green turtle.) D. 27.
        - Shell, (turtles.) D. 13.
        - Perfume. D. 28.
    - Methods of capture and transportation. E. 3.
28. LACERTILIA. (Lizards, skinks, horned-toads, chameleons, scorpions, joint-snakes, &c.)
    - Useful products: Food of Indians. D. 1.
    - Medicinal product: (Skink.) D. 30.
    - Useful traits: Destroy noxious insects.
29. OPHIDIA. (Snakes.)
    - Useful products:
        - Leather, (rattlesnakes, bull snakes.) D. 27.
        - Medicinal products, (rattlesnakes, copperheads.) D. 30.
        - Oil, (rattlesnakes.) D. 27.
    - Useful traits: Destroy vermin.
    - Injurious traits: Enemies of man, (rattlesnakes, copperheads, and moccasins.)

IV. AMPHIBIANS.

30. ANURA. (Frogs, toads, hyla, &c.)
    - Useful products:
        - Food, (frogs.) D. 1.
        - Material for physiological instruction, (frogs.)

30. ANURA—Continued.

Useful products:

Weather indicators, (hyla.)

Useful traits: Destroy noxious insects, (toads.)

31. URODELA. (Salamanders, axolotls, and menopomes.)

Useful products: Foods, aboriginal, (axolotls.)

Useful traits: Aquarium use.

Injurious traits: Enemies of young fish.

32. PROTEIDA. (River-dogs, hell-benders.)

Injurious traits: Enemies of young fish.

33. TRACHYSTOMATA. (Sirens.)

V. FISHES.

34. PEDICULATI. (Sea-bats or devil-fish, goose-fish or angler, mouse-fish, &c.)

Useful products: Baits, (goose-fish.) D 1.

Injurious traits: Enemies of aquatic birds, (goose-fish.)

35. PLECTOGNATHI. (Sun-fish, rabbit-fish, porcupine-fish, swell-fish, box-fish, trunk-fish, cow-fish, file-fish, trigger-fish.)

Useful products:

Food, (file-fish, trunk-fish.) D. 11.

Clothing, (helmets made from porcupine-fish.) D. 20.

Oils, used in medicine, (sun-fish.) D. 27.

Shagreen, (file-fish, trigger-fish.) D. 30.

36. LOPHOBRANCHII. (Sea-horse, pipe-fish.)

Useful traits: Aquarium use.

37. HEMIBRANCHII. (Snipe-fish, trumpet-fish, stickleback.)

Useful traits: Aquarium use, (sticklebacks.)

Injurious traits: Destroy eggs of other fishes.

38. TELEOCEPHALI:

HETEROSOMATA, (soles, flounders, flatfish, turbot, halibut.)

Useful products:

Foods: Fresh. D. 1.

Smoked, (halibut.) D. 2.

Pickled, (halibut.) D. 3.

Baits, D. 5.

38. TELEOCEPHALI—Continued.

ANACANTHINI, (cod, pollock, haddock, hake, ling, cusk, burbot, rockling, lance.)

Useful products:

Food: Fresh. D. 1.

Salted, wet, (cod,) (cods' sounds, tongues.) D. 3.

Salted, dry, (cod, haddock, hake.) D. 2.

Bait, (lance.) D. 5.

Isinglass, (cod, haddock, hake.) D. 24.

Leather, (N. W. coast Indians, cod.) D. 27.

Oil, (cod, haddock, hake, livers.) D. 30.

ACANTHOPTERI, (Wolf-fish, blenny, oyster-fish, toad-fish, lump-fish, sea-snail, goby, sea-robin, gurnard, sculpin, sea-raven, Norway haddock or hemdurgan, red-fish, rock cod (west coast), black-fish, or tautog, cunner or chogset, parrot-fish, viviparous-fish (west coast), surgeon-fish, angel-fish, chætodons, sword-fish, bayonet-fish, scabbard-fish, mackerel, cero, tunny, bonito, crevallé, pompano, pilot-fish, dolphin, butter-fish, weak-fish, drum, croaker, king-fish, whiting, bass, sheepshead, scup or porgy, grunts or pig-fish, black bass, sunfish, strawberry bass, rock bass, perch, groupers, striped bass or rock-fish, blue-fish, tailor, cobia, remora, barracuda.)

Useful products:

Food: Fresh. D. 1.

Salted, wet, (sword-fish, mackerel, tunnies, pompanoes, blue-fish.) D. 2.

Baits. D. 5.

Isinglass, (weak-fish, drum, &c.) D. 24.

Ornament, scales, (parrot-fish, drum.) D. 14.

Injurious traits:

Poisonous, (barracuda, dolphin, &c.)

Enemies of vessels, (sword-fish, bayonet-fish.)

Parasitic on useful fishes, (remora, toad-fish, and sea-snail, (on oysters and pectens,) &c.)

Bait-thieves, (sculpins.)

38. TELECOPHALI—Continued.

PERCESOCES. (Atherines, mullet.)

Useful products:

Food: Fresh. D. 1.

Salted, smoked, (mullet, mullet-spawn.) D. 2.

Salted, wet, (mullet.) D. 3.

Bait, (atherines.) D. 5.

Scales, (mullet.) D. 14.

SYNENTOGNATHI. (Gar-fish, flying-fish.)

Useful products:

Food: Fresh. D. 1.

Salted, smoked, (gar-fish, flying-fish.) 1.

HAPLOMI. (Blind-fish, pike, pickerel, minnows.)

Useful products:

Food, (pike, pickerel.) D. 1.

Bait, (minnows.) D. 5.

Injurious traits:

Enemies of other fishes and of aquatic birds, (pikes, pickerels.)

ISOSPONDYLI. (Capelin, oulachan, smelt, white-fish, salmon, trout, tarpum, herring, menhaden, shad, alewife or gaspereau, anchovy, &c.)

Useful products:

Food: Fresh. D. 1

Salted, (shad, salmon, white-fish, herring, &c.) D. 2.

Smoked, (herring, salmon, &c.) D. 1.

Canned, (salmon, menhaden, sardines, &c.) D. 1.

Eggs. D. 1.

Sauce, (anchovy.) D. 3.

Oil, (salmon, oulachan, white-fish, menhaden, herring.) D. 27.

Bait, (capelin.) D. 5.

Ornamental scales, (tarpum.) D. 14.

Guano, (menhaden, herring, &c.) D. 31.

Modes of culture. E. 9.

EVENTOGNATHI. (Suckers, dace, buffalo-fish, carp, tench, &c.)

Useful products:

Food. D. 1.

Bait. D. 5.

Artificial pearls. D. 29.

Modes of culture, (including domesticated species.) E. 9.

39. NEMATOGNATHI. (Cat-fish, "bull-heads," &c.)
    Useful products:
        Food. D. 1.
        Guano, (cat-fish.) D. 31.
40. APODES. (Eels, congers.)
    Useful products:
        Food. D. 1.
        Bait, eel-skins. D. 5.
        Leather, (eels.) D. 20.
41. CYCLOGANOIDEI. (Mud-fish, or amia.)
42. RHOMBOGANOIDEI. (Gar-pikes.)
    Useful products: Scales, used for arrow-tips. D. 14.
    Injurious traits: Enemies of other fish.
43. SELACHOSTOMI. (Paddle-fish, or spoon-bill)
44. CHONDROSTEI. (Sturgeons.)
    Useful products:
        Foods: Fresh. D. 1.
            Smoked. D. 1.
            Eggs, pickled, (caviare.) D. 3.
            Chorda-dorsalis, dried, (veziga.) D. 2.
            Isinglass, (sturgeon.) D. 24.
        Oil. D. 27, T.
        Scales. D. 14.
    Useful traits: Scavengers.
    Injurious traits: Said to destroy eggs of white-fish.

VI. ELASMOBRANCHIATES.

45. HOLOCEPHALI. (Chimæra, or king of the herrings.)
46. RAIÆ. (Skates, rays, "devil-fish.")
47. SQUALI. (Sharks.)
    Useful products:
        Food, (sharks, skates.) D. 1.
        Bone, (sharks.) D. 9.
        Oil, livers, (sharks, rays, &c.) D. 27.
        Shagreen, (sharks.) D. 20.
    Injurious traits: Enemies of man and fishes.

VII. MARSIPOBRANCHIATES.

48. HYPEROARTIA. (Lamprey-eels, or nine eyes.)
49. HYPEROTRETI. (Suckers, or hags.)
    Useful products: Food, (lamprey-eels.) D. 1.
    Useful traits: Scavengers, (hags.)

VIII. LEPTOCARDIANS.

50. CIRROSTOMI. (Amphioxus.)
    Modes of capture. D. 6.

IX. INSECTS.

51. HEXAPODA. (Bees, butterflies and moths, flies, beetles, bugs and lice, grasshoppers and crickets, dragon-flies and caddice flies.)
    Useful products:
        Food of aborigines.
        Honey, (bees, &c.) D. 1.
        Wax, (bees, &c.) D. 30.
        Baits, (flies, bees, dragon-flies, beetles and their larvæ, grasshoppers, &c.) D. 5, B. 45.
        Silk, (moths.) D. 8.
        Coloring material, (cochineal insect, &c.) D. 29.
        Blistering preparations, (Spanish-flies, &c.) D. 30.
        Wings used in the arts, (beetles.) D. 19.
    Useful traits:
        Puncture trees, producing galls, manna, lac, &c.
    Injurious traits:
        Injurious to vegetation, (numerous species.)
        Internal and external parasites, (flies.)
52. MYRIAPODA. (Centipedes, millipedes.)
    Useful products: Food of aborigines, (eggs.) D. 1.
    Injurious traits: Venomous, (centipedes, millipedes.)

X. ARACHNEANS.

53. ARACHNIDA. (Spiders, scorpions, mites, &c.)
    Useful products:
        Fine threads used by opticians, (spiders.)
        Silk, (spiders.)

53. ARACHNIDA—Continued.

Useful traits: Destroy noxious insects, (spiders.)

Injurious traits:

Venomous, (scorpions.)

Parasites, (mites.)

XI. ARTHROPODS.

54. CRUSTACEA. (Crabs, lobsters, shrimps, prawns, crawfish, limnoria, fish-lice, lernæans, sand and water fleas, barnacles, horseshoe crabs, &c.)

Useful products:

Foods, (fresh and canned crabs, lobsters, shrimps, prawns, crawfish, lobsters.)

Baits, (crabs, lobsters, shrimps, prawns, &c.)

Manures, (horseshoe crabs.)

Useful traits: Skeleton cleaners, (beach fleas, &c.)

Injurious traits:

Parasites on fishes and marine mammals, (barnacles, fish-lice, &c.)

Destroy earthworks, dams, &c., (crawfish.)

Destroy submerged timbers, (limnoria, &c.)

Modes of protection against injurious species. E. 4.

Methods of capture. D. 7, 31, 32.

XII. WORMS.

55. ANNELIDA. (Sipunculoids, leeches, earth-worms, serpulæ, sea-worms, &c.)

Useful products:

Food of aborigines, (earth-worms.) D. 1, 2.

Baits, (earth-worms, sea-worms.) D. 5.

Useful traits:

Used in surgery, (leeches.)

Used as barometers, (leeches.)

Injurious traits: External parasites of animals, (leeches.)

Methods of culture, (leeches.) E. 11.

56. SCOLECIDA. (Tape-worms and flukes, planarians, nemerteans, trichinæ, thread-worms, rotifers, &c.)

Injurious traits: Internal parasites, (numerous species.)

XIII. MOLLUSKS.

57. CEPHALOPODA. (Octopus, nautilus, argonauts, calamaries or squids.)

Useful products:

Food, (squids and their eggs.) D. 1.

Bait, fresh and salted, (octopus, squids.) D. 1, 5.

Ink, sepia, (sepias.) D. 29.

"Bone," used as food for animals. D. 5.

"Bone," used in arts and manufactures. D. 19.

58. GASTROPODA. (Land-snails, sea-snails, whelks, limpets, &c.)

Useful products:

Food, (numerous species.) D. 1, 2.

Bait, (limpets, &c.) D. 5.

Nacre, (top-shells, ear-shells, &c.) D. 15.

Shell used in arts and manufactures. D. 16.

Useful traits:

Carrion-feeders, (strombus and other siphonated genera.)

Food of useful animals.

Injurious traits:

Predatory on other mollusks, (murex, buccinum, natica, &c.)

Injurious to vegetation.

59. CONCHIFERA. (Ordinary bivalve shells.[1])

Useful products:

Food, fresh, dried, and pickled, (numerous species.) D. 1, 2, 3.

Baits, (clams, mussels, &c.) D. 5.

Pearls and nacre, (river-mussels, pearl-oysters, &c.) D. 15.

Shell used in arts and manufactures. D. 16.

Injurious traits: Borers in wood and stone, (ship-worms, pholas, gastrochæna, date, shells, saxicava, ungulina, &c.)

XIV. RADIATES.

60. ECHINODERMATA. (Sea-cucumber, sea-urchins, star-fishes, ophiurans.)

Useful products:

Food, fresh, (sea-urchins and their eggs.) D. 1.

Food, dried, (bêches le mer.)

---

[1] *Tunicata, brachiopoda,* and *bryzoa* are omitted, on account of their very remote usefulness.

60. ECHINODERMATA—Continued.

Injurious traits:

Burrowers, (various echinoids.)

Destroyers of useful mollusks.

61. CŒLENTERATA. (Acalephs, polyps, &c.)

Useful products: Coral, various species of polyps.) D. 17.

Injurious traits: Clog seines, weirs, and fishing-lines, (acalephs.)

XV. PROTOZOANS.

62. RHIZOPODA. (Sponges and foraminifera.)

Useful products:

Food, "mountain meal," (foraminifera.) D. 1

Infusorial earths, (foraminifera.) D. 18.

Sponges, used in arts and manufactures. D. 26.

SECTION B.

(THE CHASE AND THE FISHERIES.)

MEANS OF PURSUIT AND CAPTURE.

I. HAND IMPLEMENTS OR TOOLS.

* *For striking.*

1. CLUBS:
   - *a.* Unarmed clubs:
     - Salmon-clubs, used by the Indians of the Northwest coast.
     - Other fishing-clubs.
     - Hunting-clubs.
   - *b.* Armed clubs:
     - Stone-headed clubs.
     - Clubs, armed with teeth or bone points.
     - Clubs, armed with metal points.
2. SLUNG-WEIGHTS:
   - *a.* Slung-stones.
   - *b.* Slung-shot.
   - *c.* ("Morning stars.")
   - *d.* ("Flails.")

** *For cutting.*

3. KNIVES:
   - *a.* Straight knives:
     - Hunting-dirks and daggers.
     - Hunting-knives, scalp-knives, &c.
     - Blubber-knives, aboriginal and recent.
     - Boarding-knives used by whalemen.
     - Whaleman's boat-knives.
     - Bowie-knives.
     - Flaying-knives, aboriginal and recent.
     - Splitting-knives.
     - Heading-knives.
     - Sailors' and fishermen's sheath-knives.
     - Hunters' sheath-knives.
     - Slivering-knives, used by fishermen.
     - Oyster-knives.

3. KNIVES—Continued.
    a. Straight knives:
        Mackerel rimmers or fatting knives.
        (Swords, including the various forms incidentally used in hunting, sabers, cutlasses, machétes, creases, &c.)
        Stone and bone knives, used by Indians and Eskimos.
        Skin scrapers and parers, used in preparing leather.
    b. Clasp-knives:
        Sailors' clasp-knives.
        Hunters' clasp-knives.
        Clasp-dirks.
        Jockey knives.

4. AXES:
    a. Axes, proper:
        Tomahawks.
        Hatchets.
        Whaleman's boat-hatchets.
        Cleavers.
        Axes, used by fishermen and hunters.
        Head-axes for whalemen.
    b. Cutting-spades:
        Whale-spades:
            Cutting-spades.
            Throat-spades, flat and round shank.
            Wide spades.
            Half-round spades.
            Head-spades.
        Blubber-mincing knives.
        Chopping-knives.

*** *For thrusting.*

5. THRUSTING SPEARS AND PRODS:
    a. Fishing-lances.
        Whale-lances.
        Whaleman's boat-spades, thick and thin.
        Seal-lances.
        Fish-lances.
    b. Hunting-spears.
    c. Bayonets.
    d. Prodding-awls, used in piercing the base of the brain in killing fish for the table.

II. IMPLEMENTS FOR SEIZURE OF OBJECT.

* *Scooping-instruments.*

6. SCOOPS.

† *For hand-use.*

a. Shovels:

Clam-shovels.

Trowels used in taking burrowing shore animals.

Hand-scoops.

b. Hand-dredges, used in collecting mollusks.

c. Pile-scrapers.

†† *For use with sounding-lines.*

d. Armed leads:

Common "deep-sea lead."

Deep-sea-sounding apparatus.

e. Cup-leads.

f. Scoop sounding-machines.

** *Grasping-hooks.*

7. HOOKED INSTRUMENTS. (Those used with a single motion, that of hooking:)

a. Single-pointed hooks:

Gaff-hooks.

Boat-hooks.

Jigs.

Rabbit and squirrel hooks, used by the Ute Indians.

Snake-hooks.

Clam-hooks.

Hoes and picks used in gathering shell-fish.

Forks used in handling salted and dried fish.

Whalemen's hooks:

Blubber-hooks.

Blubber-forks.

Junk-hooks.

Lance-hooks.

Can-hooks.

7. HOOKED INSTRUMENTS—Continued.

b. Many-pointed hooks:

Grappling-irons.
Lip hooks or grapnels, used by whalers.
Toggles, used by whalers.
Oyster-rakes.
Clam-rakes.
Oulachan rakes or spears.
Squid-jigs.

c. Twisting-rods, used in drawing small mammals from their burrows.

8. BARBED IMPLEMENTS. (Those used with two motions, the first that of thrusting:)

a. Spears with fixed heads:

Harpoons.
One-flued harpoons.
Two-flued harpoons.
Toggle-harpoons.
Harpoon-bullets. (See under 23.)
Gun-harpoons.
Other whaleman's "craft."
Barbed spears, (with single point.)
Grains, (with two prongs.)
Gigs.
Bird-spears.
Otter-spears.
Sea-otter spears.
Seal-spears.
Walrus-spears.
Eel-spears.
Flounder-spears.
Sturgeon-spears, (west coast.)
Octopus-spears.
Crab-spears, used in Rhode Island.

b. Spears with detachable heads:

Lily-irons.
Dolphin-irons.
Indian harpoons of shell and iron.
Eskimo harpoons of stone, bone, and iron.

8. BARBED INSTRUMENTS—Continued.

b. Spears with detachable heads:

Indian fish-harpoons.

Other fish-harpoons.

(For accessory apparatus, see under 29.)

9. TONGS, &c.

† *For hand use.*

a. Tongs (with two handles:)

Oyster-tongs.

Oyster-rakes.

b. "Nippers," (with cord and handle.)

Snake-tongs.

Sponge-tongs.

Coral-tongs.

†† *For use with sounding-lines.*

c. "Clamms" for deep-sea soundings, (forceps closed by a weight.)

(Ross's "Deep-sea clamms.")

(Bull-dog sounding-machine.).

*** *Grasping-lines.*

10. NOOSES.

† *Stationary nooses.*

a. Jerk-snares:

Bird-snares.

Fish snares, of wire, gut, hair, &c.

†† *Thrown nooses.*

b. Lariats and lassos:

Lariats with rope noose, made from hair, hemp, and rawhide.

Lariats with metal noose.

(Chilian bird-lariat.)

11. LOADED LINES. (Bolas.)

a. Bird-slings, used by Eskimos.

b. Bolas, with one or several weights.)

**** *Entangling lines.*

12. TANGLES.

a. Tangles:

Swab-tangles.

(Dredge-tangles, used by English collectors.)

Harrow-tangles.

Wheel-tangles.

III. MISSILES.

* *Simple missiles, (those propelled by the unaided arm.)*

13. HURLED WEIGHTS.
    - *a.* Stones and discs thrown by the hand.
    - *b.* Weights dropped from an elevation, (dead-falls, not automatic.)
14. HURLED STICKS.
    - *a.* Straight sticks:
        - Clubs used as missiles.
    - *b.* Curved sticks:
        - Throw-sticks, used by the Moqui Indians of New Mexico in hunting rabbits.
        - (Boomerangs.)
15. HURLED SPEARS.
    - *a.* Darts and lances.

** *Centrifugal missiles. (Propelling power augmented by an artificial increase of the length of the arm.)*

16. SLINGS AND SPEARS THROWN BY STRAPS.
    - *a.* Slings.
    - *b.* Spears, with straps used in throwing them.
17. MISSILES PROPELLED BY "THROWING-STICKS."
    - *a.* Spears with throwing-sticks, used by Eskimos:
        - Series of throwing or darting sticks.

*** *Missiles propelled by a spring.—‡ Spring consisting of bent rod.*

18. BOWS AND ARROWS.
    - *a.* Bows:
        - Simple bows.
        - (Cross-bows.)
        - (Ballistas.)
    - *b.* Arrows:
        - Lance-arrows.
        - Harpoon-arrows, used in fishing.
        - Blunt or club arrows, used in killing birds.
    - *c.* Accessories of bows and arrows:
        - Holders.
        - Quivers.
        - Arrow-head pouches.

18. BOWS AND ARROWS—Continued.

   d. Implements of manufacture:

      Flint-chipping apparatus.

      Arrow-head sharpeners.

      Shaft-gauges.

      Cord-twisting apparatus.

      Shaft-polishers.

      Glue-sticks, used in fastening head of arrow.

‡‡ *Spring consisting of elastic cord.*

19. INDIA-RUBBER SLINGS.

   a. Pea-shooters, used in killing birds.

‡‡‡ *Spring consisting of metallic helix.*

20. SPRING-GUNS.

   a. Spring-guns.

**** *Missiles propelled by the compression of air or water.*

21. AIR-GUNS.

   a. Blow-guns, (missile propelled by the breath:)

      Blow-guns carrying arrows.

      Blow-guns carrying balls.

   b. Piston air-guns.

   c. Reservoir air-guns:

      Air-guns.

      Air-gun canes.

22. WATER-GUNS.

   a. Syringe-guns:

      Humming-bird guns.

***** *Fire-arms.*

23. GUNS AND PISTOLS.

   a. Muzzle-loading arms:

      With smooth bores:

         Muskets.

         Fowling-pieçes.

         Cane-guns.

         Pistols:

            Single-barreled pistols.

            Revolvers.

      With grooved bores: Rifles.

         Rifle-muskets.

         Rifle-carbines.

         Pistols.

23. GUNS AND PISTOLS—Continued.

b. Breech-loading arms:

With smooth bores:

Fowling-pieces.

Pistols.

With rifled bores:

Muskets.

Hunting rifles.

Carbines:

Single-barreled carbines.

Revolving carbines.

Pistols:

Pistols.

Revolvers.

c. Whaling-guns:

Bomb lance and gun.

Harpoon ball and gun.

Harpoon-gun.

Harpoon bomb-lance gun.

24. (ACCESSORY.) AMMUNITION AND ITS PREPAR

a. Explosives:

Gunpowder.

Gun-cotton.

Percussion powder:

Caps.

Needle percussion.

Primers.

Wood powder.

Dynamite or giant-powder.

Nitroglycerine.

Dualine.

Lithofracteur.

Colonia powder.

Other explosives.

b. Missiles:

Bullets.

(Accessory) bullet-molds.

Shot.

(Accessory) methods of manufacturing shot.

24. (ACCESSORY.) AMMUNITION, &c.—Continued.

b. Missiles:

Explosive bullets, shells, &c.:

Bomb-lance.

Meigs's shells.

c. Wadding:

Bulk wadding.

Prepared wads.

(Accessory) wad-cutters.

d. Ammunition-measures:

Measures.

Shot-measures. } Attached to pouches and separate.
Powder-measures. }

Weighing-scales.

e. Prepared ammunition:

Cartridges:

Ball-cartridges.

Shot-cartridges.

Wire-cartridges.

(Accessory) paper-shells.

(Accessory) metallic shells.

f. Methods of preparing cartridges:

Loaders.

Crimpers.

Cappers.

25. ACCESSORIES OF LOADING, CLEANING AND REPAIRING, SIGHTING, AND TESTING FIRE-ARMS.

a. Instruments for cleaning, loading, &c.:

Rammers.

Swabs.

Charge-drawers, "worms."

b. Sights, &c.:

Muzzle-sights:

Plain sights.

Slit-sights.

Globe-sights.

Peep-sights.

Breech-sights:

Plain sights.

Graduating sights.

25. ACCESSORIES OF LOADING, &c.—Continued.

b. Sights, &c.:
- Telescope-sights.
- Levels, attached to guns
- Wind-gauges.

c. Targets:
- Practice-targets.
- "Gyro-trap" targets.
- Pigeon-traps and accessories of pigeon-shooting.

d. Recoil-checks.

26. FOR CARRYING ARMS AND AMMUNITION.

a. Ammunition-holders:
- Powder-holders:
  - Horns.
  - Flasks.
  - Canisters.
- Shot-holders:
  - Pouches.
  - Belts.
- Cartridge-holders:
  - Pouches.
  - Boxes.
  - Belts.
  - Vests.
- Cap-holders:
  - Pouches.
  - Boxes.
  - Cap-straps, used by Indians.

b. Weapon-holders:
- Slings for arms:
  - Shoulder-slings.
  - Saddle-slings.
  - Holsters.
- Belts:
  - Pistol-belts.
- Racks and cases:
  - Gun-racks.
  - Gun-cases.

IV. BAITED HOOKS. ANGLING-TACKLE.

27. HOOKS WITH MOVABLE LINES.
  - *a*. Tackle for surface-fishing:
    - Fly-fishing tackle.
    - Salmon-tackle.
    - Trout-tackle.
    - Black-bass tackle.
    - Shad-tackle.
    - Trolling-tackle:
      - Trolling-tackle.
      - Whiffing-tackle.
      - Drailing-tackle.
      - Gangs of hooks for minnow-bait.
    - Surf-tackle for throwing and hauling:
      - Striped-bass tackle.
      - Redfish or bass tackle.
      - Bluefish tackle.
    - Tide-drailing tackle:
      - Pasque and cuttyhunk bass-tackle.
  - *b*. Tackle for fishing below the surface:
    - Short hand-gear:
      - Mackerel-gear.
    - Deep-sea gear:
      - Cod-gear.
      - Halibut-gear.
      - Flounder-gear.
      - Shark-gear.
      - Tautog-gear.
      - Other bottom-gear.
    - Bobs:
      - Eel-bobs.
28. HOOKS, WITH STATIONARY LINES.—SET TACKLE.
  - *a*. Surface lines:
    - Spilliards, or floating-trawl lines.
  - *b*. Bottom-set lines:
    - Trawl-lines, or bull-tows.

29. (ACCESSORY.) PARTS AND ACCESSORIES OF ANGLING-APPARATUS AND OF HARPOON AND SEINE LINES.

a. Hooks, including a full series of unmounted hooks, of recent and aboriginal manufacture.

Plain hooks:

- Fly-hooks.
- Trout-hooks.
- Salmon-hooks.
- Cod and halibut hooks.
- Hooks for general use.
- Bass-hooks.

Jigs and drails:

- Mackerel-jigs.
- Blue-fish drails of bone and metal of the various patterns, Newport, Noank, Providence, Provincetown, &c.
- Block Island drails.
- Pearl-squids of various patterns.
- Bone-squids.
- Metal-squids.
- Petticoat-squids of flannel, &c.

Spoon-baits, plain and fluted:

- Bass-spoons.
- Pickerel-spoons.
- Trout-spoons.
- Blue-fish spoons.
- Other trolling-spoons.

Artificial flies on hooks:

- Salmon-flies for each month.
- Trout-flies for each month.
- (Accessory.) Fly-books.

b. Lines, (twisted and plaited:)

- Silk-lines.
- Grass-lines.
- Linen-lines.
- Cotton-lines.
- Cotton-hemp lines.
- Bark-lines.
- Manila-lines.
- Hide-lines.

29. (ACCESSORY.) ANGLING-APPARATUS, &c.—Continued.

b. Lines, (twisted and plated:)

Gut-lines.

Lines made from sea-weed, (*Nereocystis Lütkeana*,) and used by natives of Alaska.

(Linesof sea-weed, (*Chorda filum*,) used similarly in Scotland.)

(Accessory.) Apparatus for twisting lines.

c. Snoods, leaders, and traces:

"Cat-gut," (sheep,) snoods, and leaders.

Silk-worm-gut snoods.

Salmon-gut snoods.

Flax-snoods.

Gimp-snoods.

Wire-snoods.

"Sid-straps."

d. Whalers' chains and lines:

Head chains and ropes.

Fin-chains.

Fluke chains and rings and ropes.

Head pike and ring.

(Accessory.) Blocks, pendants, cutting-blocks, &c.

e. Sinkers:

Boat-shaped sinkers, plain and shearing.

Pipe-lead sinkers.

Bullet-sinkers.

Plummet-sinkers, sugar-loaf, pear-shaped, and double-taper.

Banker-sinkers.

Seine-sinkers, of chain, lead balls, lead rings, stone, &c.

(Accessory.) Molds for sinkers.

Jig-molds.

Other sinker-molds.

f. Spreaders:

Chopsticks.

One-armed chopsticks, or "revolving booms."

g. Floats:

Line-floats of wood, cork, and quill.

Harpoon-floats of bladder, inflated skir, and wood.

Seine-floats of cork, wood, glass, and rubber-tubing.

Keg and other floats for lobster-pots, gill-nets, &c.

Whale-line drag.

29. (ACCESSORY.) ANGLING-APPARATUS, &c.—Continued.

h. Reels:

Simple reels for fly-fishing, with and without check.
Multiplying reels for bass-fishing, with and without check.
Other multiplying reels.
Gunwale-winches.
Dredge-line rollers.
Trawl-line rollers.
Seine-windlasses.

i. Line-holders:

Winders.
Spools.
Whaleman's line-tub.
Tubs for trawl-lines.
Seine-reels.

k. Rods:

Straight rods, of cane, wood, whalebone, &c.:

Salmon-rods.
Trout-rods.
Bass-rods.
Pickerel-rods.
Other rods.
Folding-rods.
Tips of rubber, whalebone, &c.
Tell-tales, used in trolling.
Tell-tales for fishing under the ice.
(Accessory) cases for rods and rod-tops.

l. Swivels:

Box-swivels.
Hook-swivels.
Pot-gauge swivel.
Cod-line swivels.
Trawl buoy-rope swivels.

m. Clearing-rings.

n. Disgorgers.

V. NETS.

30. ENTANGLING-NETS.

a. Meshing-nets, (entangling in meshes:)

‡ BARRIER-NETS.

Rabbit-nets, used by Indians of the Southwest.

Bird mesh-nets.

Gill-nets, used in great lakes.

‡ DRIFT-NETS.

† *Those drifting across the tide.*

Shad gill-nets, used in southern rivers.

Bass gill-nets.

Salmon gill-nets.

Mullet gill-nets.

†† *Those drifting along the tide.*

Mackerel gill-nets.

Herring gill-nets.

b. Pocket-nets, (entangling in pockets:)

Trammel-nets.

31. ENCIRCLING-NETS.

a. Seines:

Seal-seines.

Manatee-seines.

Shad-seines.

Mullet-seines.

Menhaden-seines.

Bass-seines.

Blue-fish seines.

Capelin-seines.

Herring-seines

Cod-seines.

Lance-bunts.

Baird collecting-seines.

Bait-seines.

"Fly-tail" seines of North Carolina.

b. Hoop-nets:

Handle, or dip-nets:

Bull-nets, (worked with ropes and blocks.)

31. ENCIRCLING-NETS—Continued.

b. Hoop-nets:

Handle or dip-nets:

Scoop-nets, (herring-nets, pound-scoops, car-scoops, &c.)

Landing-nets.

Eskimo auk-nets.

Baited hoop-nets:

Crab-nets.

c. Trailing-nets:

Trawls:

Beam-trawl.

(Otter-trawl.)

Dredges:

Flange, or ordinary dredge.

Rake-dredge.

Oyster-scraper.

(Coral-dredge.)

Towing-nets:

Surface tow-nets.

d. Folding or jerk nets:

Purse-nets:

Mackerel purse-seines, (pursed by weight.)

Menhaden purse-seines, (pursed by hand-ropes.)

Cast-nets:

Mullet cast-nets.

Pompano cast-nets.

Bait cast-nets.

Clap-nets for birds.

Rabbit-spring nets.

Spring-weirs, (St. Lawrence.)

Sieve-traps, (for birds.)

e. (Accessory.) Parts of nets and apparatus for manufacture:

Raw material of nets.

Babiche. (See under D. 20.)

Netting-fibre.

Netting-twine.

Netting-needles.

Mesh-needles.

Hanging-needles.

Eskimo netting-needles.

VI. TRAPS.

32. PEN-TRAPS.

a. Pocket-traps:

Pitfalls:

Pits, covered.

Barrel-traps.

Jar mole-traps.

"Rabbit-tipe," used in England.

Salmon-baskets, (Columbia River.)

Salmon-weirs, (Upper Columbia River.)

River-weirs, with pockets:

Eel-traps.

Fish-slides:

Shad-slides, used in the rivers of North Carolina.

b. Labyrinth-traps:

Corrals.

Turkey-traps.

Weirs, or pounds:

Heart-pound.

Salmon-weir.

Virginia Indian weir, (figured by DeBry.)

Salmon hook-gill-net of the Saint Lawrence.

Funnel-traps:

Fish-pots.

Lobster-pots.

Eel-weirs, (with leaders.)

Eel-pots, (without leaders.)

Barrel-pots, for eels.

West India wicker fish-pots.

Set-nets.

Fykes, (set-nets with leaders.)

Bass-traps.

c. Door-traps:

† *Closed by the falling of a door.*

Box-traps.

Rabbit-traps, (figure 4.)

Brick traps, (figure 4.)

32. PEN-TRAPS—Continued.

c. Door-traps:

Box-traps:

Musquash traps, with hanging doors.

Rabbit-traps, for mouth of burrows.

Self-setting box-traps.

Double box-traps.

Spring-door traps.

†† *Closed by falling of whole trap.*

Bowl-traps.

Cob-house bird-traps.

Pigeon-nets.

††† *Closed by falling of tide.*

Bar-weirs.

d. Sheaf-traps:

Sheaf-traps, (New York Harbor.)

33. CLUTCHING-TRAPS.

a. Noose-traps:

Snares:

Footpath-snares.

Barrier-snares.

Springes.

"Round mouse-traps."

b. Jawed traps:

"Steel traps:"

Newhouse traps:

No. 0. Rat-trap.

No. 1. Muskrat trap.

No. 1½. Mink-trap.

No. 2. Fox-trap.

No. 3. Otter-trap.

No. 4. Beaver-trap.

No. 4½. Deer-trap.

No. 5. Small bear-trap.

No. 6. Great bear-trap.

Spring bird-nets.

(French bird-trap.)

34. FALL-TRAPS.
   - *a.* Crushing-traps:
     - Deadfalls.
     - Figure-four traps.
   - *b.* Piercing-traps:
     - Spear-falls.
     - Mole-traps.
     - Harpoon-traps.
   - *c.* Spring-hooks:
     - Pickerel-hooks.

35. MISSILE-TRAPS.
   - *a.* Cross-bow traps.
   - *b.* Spring-guns.

36. ADHESIVE PREPARATIONS.
   - *a.* Bird-lime, &c.
   - *b.* Hoods, boots, &c.

VII. APPARATUS FOR WHOLESALE DESTRUCTION.

37. POISONS.
   - *a.* Food poisons:
     - Phosphorus poisons.
     - Strychnine.
     - Arsenic.
     - Corrosive sublimate.
     - Cyanide of potassium.
     - Opium poisons.
   - *b.* Blood poison: Woorara

38. ASPHYXIATORS.
   - *a.* Apparatus for smoking-out.
   - *b.* (Apparatus for suffocating with fumes of sulphur.)
   - *c.* Apparatus for drowning-out.

39. TORPEDOES.

39½. STOMACH-SPRINGS.
   - *a.* Eskimo whalebone springs, used in killing bears.

VIII. HUNTING-ANIMALS.

40. HUNTING-MAMMALS.
   - *a.* Dogs.
   - *b.* Hunting-leopard. (*Cynailurus jubatus.*)

40. HUNTING-MAMMALS—Continued.

c. Weasels and ferrets.

d. Otters.

41. ACCESSORIES TO HUNTING-DOGS.

a. Dog-whips.

b. Dog-whistles.

c. Dog-collars.

d. Dog-food.

e. Dog-carts.

f. Dog-muzzles.

42. HUNTING-BIRDS.

a. Falcons.

b. Owls.

c. Cormorants, (*Carbo sinensis*, used in fishing in China.)

43. ACCESSORY TO HUNTING-BIRDS.

a. Hoods.

b. Perches.

c. Cormorant-collars.

44. HUNTING-FISHES.

a. Remora, used in West Indies and Australia.

IX. DECOYS AND DISGUISES.

45. BAITS.

a. Natural baits:

Flies and other insects. (This should include a collection of those insects which, as the favorite food of fishes, are imitated in making artificial flies.)

Worms.

Mollusks.

Salted baits, (prepared.)

Menhaden.

Herring.

Squids.

Clams, long.

Clams, hen.

Pea-roe of cod, (used in French sardine-fisheries, and largely exported.)

Grasshopper paste, used as a substitute for pea-roe.

Tolling baits, "stosh," &c.

45. BAITS—Continued.

a. Natural baits:

(Accessory) methods of preparing baits:

Bait-cutters.

Bait-mills.

Bait-ladles.

Wheelbarrows for bait-clams, (Nantucket.)

Bait boxes and cans.

Bait-needles.

b. Artificial baits:

Trolling-spoons.

Spinners.

Squids and jigs.

"Bobs," used in southern waters.

Artificial flies.

c. Accessory to *b*:

a. Fly-books.

b. Raw materials for making artificial flies.

c. Tools for making artificial flies.

d. Pastes.

46. DECOYS.

a. Scent-decoys.

b. Sound-decoys:

Animal calls, whistles, &c.

Bird-calls.

c. Sight-decoys:

Living decoy animals and birds.

Decoy-dogs, used in hunting ducks.

Stool-pigeons.

Tame decoy-ducks.

Tame decoy-brants.

Imitations of animals and birds:

Decoy swimming-birds.

Decoy-waders.

Imitations of fishes:

Lure-fish used in taking Mackinaw trout.

Blanket-decoys, (for antelopes.)

Lanterns and other apparatus for fire-hunting and fishing,

Lanterns for still-hunting.

46. DECOYS—Continued.

c. Sight-decoys:

Lanterns for weequashing, or fire-fishing, for eels.

Jack-lanterns for fishing.

47. COVERS.

a. Movable covers:

Masks:

Deer heads and antelope heads.

Movable copses.

Covers for hunter.

Covers for boats.

b. Stationary covers:

Hunting-lodges.

X. PURSUIT, ITS METHODS AND APPLIANCES.

48. METHODS OF TRANSPORTATION.

a. Personal aids:

Snow-shoes.

Skates.

Alpenstocks and staves.

Portable bridges.

b. Animal equipments:

Harness:

Horse-trappings.

Dog-harness.

Girths, sinches.

Bits, cabrestos, spurs.

Saddles:

Riding-saddles.

Pack-saddles.

Aparejos.

Riding-pads, (for buffalo-hunting.)

Furpack-saddle, (Hudson's Bay Territory.)

Vehicles:

Deer-sledges.

Dog-sledges.

Wagons.

Dog-carts.

Fish-carts, used in Nantucket.

48. METHODS OF TRANSPORTATION—Continued.

c. Boats:

Hunting-boats, fishing-boats:

- Birch canoes.
- Canoes used by Indians of the northwest coast in whaling.
- Kyaks or bidarkas.
- Umiaks or bidarras.
- Indian raft-boats.
- Launches.
- Dug-outs.
- Portable (paper and canvas) boats.
- Duck-boats.
- Scows.
- Oyster-boats.
- Whale-boats.
- Seine-boats, (sea use.)
- Seine-boats of the lakes.
- Potomac seine-boats.
- Dorys, sharpies, and dingies.
- Pound-boats of the lakes.
- Italian fishing-boats, (California.)
- Pinkies, (Martha's Vineyard.)
- Adirondack boats.
- Alexandria Bay boats.
- Surf-boats.
- Whitehall boats.
- Oyster-canoes.
- Ducking-boats.
- Cat-rigged fishing-boats.
- Mackerel-smacks.
- Oyster-smacks.
- Menhaden-smacks.
- Menhaden-carryaways.
- Bank cod-smacks.
- Smacks with wells, used near the coast.
- Smacks employed in fish-trade.
- Whale-ships.
- Sealers.

48. METHODS OF TRANSPORTATION—Continued.

c. Boats:

Herring-boats.

Mackinaw boats.
Huron boats.
Norwegian boats.
Pound-boats.
} Used in the Great Lake fisheries.

Oyster-pungies, (canoe and square-sterned,) employed on the Chesapeake.

Oyster police-boats.

Steamers:

Mackerel-steamers.

Menhaden steam-mills.

Lake gill-net steamer.

Whale-steamers.

Sealing-steamers, &c.

Accessory to fishing-vessels:

Rigging, masts, sails, cordage, pulleys, sockets.

Anchors, killicks, chains.

Sail-needles, palms, fids, marline-spikes.

Oar-locks, chocks, oar-rests.

Stepping-irons for whale-boats.

Crotches and oar-rests.

Paddles and oars.

Rudder-heads, wheels, tillers, &c.

Fog-horns, trumpets, drums, &c.

Cabin, blubber room, cooks' and binnacle lamps and jacket-lamps, signal, binnacle, and common lanterns.

Compasses, barometers, &c.

Astronomical instruments, sextants, quadrants, chronometers, hour and log glasses.

49. CAMP-OUTFIT.

a. Shelter:

Lodges.

Tents.

Hunting-camps.

Hunters' houses.

Fishing-houses.

49. CAMP-OUTFIT—Continued.

b. Furniture:

Hammocks.

Beds, couches, stretchers, and lounges.

Blankets, rubber and Mackinaw, and fur robes.

Fuel.

Apparatus for kindling fire.

Lamps and lanterns.

Tools.

c. Commissary supplies:

Cooking-apparatus, kettles, and stoves.

Table-furniture.

Preserved meats, &c.

50. PERSONAL EQUIPMENTS.

a. Clothing:

Hunting-suits.

Cloth-suits.

Skin-suits.

Water-proof suits.

Oil-skin suits.

Boots, moccasins, leggings.

Water-proof boots.

Wading boots and stockings.

Riding-boots.

Moccasins.

Leggings.

Hats and caps.

Protection from insects:

Nets for beds and for face.

Ointments, (such as tar and sweet-oil.)

Smudges, (such as pyrethrum powder.)

Shields, breastplates, and defensive armor.

b. Trappings:

Belts.

Cross-belts.

Game-bags.

Game and fish baskets and slings.

Wallets for lines and other tackle.

50. PERSONAL EQUIPMENTS—Continued.

c. Optical instruments, &c.:

Snow-goggles.

Telescopes.

Field-glasses, &c.

Water-telescopes.

d. Medical outfit:

Medicine-chests.

Hunters' and fishermen's flasks.

e. Artificial lights:

Lanterns for camp and ship use.

Torches.

SECTION C.

METHODS OF PREPARATION.

I. PREPARATION AND PRESERVATION OF FOOD

1. PRESERVATION DURING LIFE, (see under E, 3.)
2. PRESERVATION OF FRESH MEATS.
    - *a.* Refrigerators:
        - Ice-boxes and refrigerators.
        - Refrigerator-cars.
        - (Accessory.) The ice-trade:
            - Ice cutting and handling apparatus.
            - Methods of manufacturing artificial ice.
            - Ice-houses.
    - *b.* Other accessories of preservation:
        - Meat-hooks.
        - Skewers, &c.
        - Carving-tools.
3. PRESERVATION BY DRYING.
    - *a.* Sun-drying apparatus:
        - Beach dryers.
        - Flake-drying:
            - Newfoundland flakes.
            - Massachusetts flakes.
            - Covers for fish-drying.
    - *b.* Smoke-drying apparatus:
        - Herring smoke-houses.
        - Halibut smoke-houses.
        - Salmon smoke-houses.
        - Sturgeon smoke-houses.
        - Aboriginal drying-houses.
        - Methods of drying haliotis, used by the Indians of California.

4. PRESERVATION BY CANNING AND PICKLING.

a. Salting fish:

Knives, (see under B, 2.)

Scaling-apparatus.

Tables, tubs, &c.

Barrels.

(Accessory.) Salt:

Specimens of the salts used in preserving fish.

Model of salt-mills used on Cape Cod in former days.

b. Canning meats:

Model of salmon-canning establishment.

Model of sardine-factory.

(Accessory.) Cotton-oil, and its manufacture.

Model of lobster-canning factory.

Model of oyster-canning factory.

5. PREPARATION OF BAITS.

a. Bait-mills, knives, choppers, &c., (see under B, 2 and 3.)

b. Bait tubs, vats, &c.

II. MANUFACTURE OF TEXTILE FABRICS, FELT AND STUFFINGS.

6. PREPARATION OF WOOL AND HAIR OF MAMMALS.

a. Preparation of wool cloths:

Washing.

Shearing.

Stapling or assorting.

Scouring.

Combing, carding, and plucking.

Spinning and reeling.

Weaving.

Fulling and teazling.

Cropping.

Pressing.

b. Weaving worsted cloths.

c. Felting and the hat manufacture:

Bowing.

Pressing.

Stopping.

6. PREPARATION OF WOOL, &c.—Continued.

c. Felting and the hat manufacture:

Rolling-off.

Shaping.

d. Preparation of curled hair for stuffings.

7. PREPARATION OF WHALEBONE.

a. Preparation of stuffings.

8. PREPARATION OF FEATHERS.

a. Preparation of down for stuffings.

b. Preparation of feather fabrics.

c. Preparation of "brillantine."

d. Preparation of or flocking for wall-paper, from refuse quills.

e. Preparation of fibres for manufacture of plush carpets.

9. PREPARATION OF SILK OF INSECTS.

a. Preparation of silk of silk-worms:

Boiling the cocoons.

Reeling.

Spinning.

Dyeing.

Weaving.

10. PREPARATION OF SOFT PARTS OF OTHER INVERTEBRATES.

a. Preparation of silk from byssus of *Pinna.*

b. Preparation of sponge stuffing.

III. PREPARATION OF THE SKIN AND ITS APPENDAGES.

11. CURRYING OF LEATHER.

a. Processes of currying:

Dipping.

Graining.

Scraping.

Dressing.

b. Implements employed by curriers:

"Head-knives."

"Pommels."

"Stretching-irons."

"Round-knives."

"Cleaners."

11. CURRYING OF LEATHER—Continued.
 b. Implements employed by curriers:
  "Maces."
  "Horses," or trestles.
  "Dressers."
  "Treading-hurdles."
 c. Eskimo and Indian currying methods and implements.
 d. Methods of dressing gut and sinew.

12. LEATHER-DRESSING.
 a. Processes of tanning leather:
  Soaking.
  Liming.
  Tanning.
 b. Processes of tawing or oil-dressing leather:
  Soaking.
  Liming.
  Oiling.
 c. Apparatus of leather-dressing, recent and aboriginal.

13. FUR-DRESSING.
 a. Processes of fur-dressing:
  Currying. (See under 12.)
  Scouring.
  Tanning.
  Lustering.
  Plucking and dyeing.

14. FEATHER-DRESSING.
 a. Method of preparing ornamental feathers:
  Scouring.
  Bleaching.
  Washing.
  Azuring.
  Sulphuring.
  Scraping.
  Dyeing.
 b. (Art of plumagery.)

15. MANUFACTURE OF QUILL ARTICLES.
 a. Manufacture of quills for pens:
  Sand-bath drying and steaming.
  Polishing.

15. MANUFACTURE OF QUILL ARTICLES—Continued.
    - *a.* Manufacture of quills for pens:
        - Dyeing.
        - Shaping.
    - *b.* Manufacture of tooth-picks.
    - *c.* Manufacture of floats and other articles.
    - *d.* Manufacture of quill brush-bristles.
16. HAIR AND WOOL WORK.

IV. PREPARATION OF HARD TISSUES.

17. IVORY CUTTING AND CARVING.
    - *a.* Manufacture of handles, trinkets, billiard-balls, &c.:
        - Turning and sawing.
        - Polishing.
        - Bleaching.
    - *b.* Manufacture of organ and piano keys:
        - Sawing.
        - Strip-sawing.
        - Polishing.
        - Bleaching, &c.
    - *c.* Other processes.
18. PREPARATION OF HORN AND HOOF.
    - *a.* Steaming.
    - *b.* Pressing.
19. PREPARATION OF WHALEBONE.
    - *a.* Cutting and other processes.
    - *b.* Manufacture of whip-makers' stock and whips.
    - *c.* Manufacture of umbrella-maker's bone.
    - *d.* Manufacture of ribbon-weaver's bone.
    - *e.* Manufacture of hat and bonnet maker's bone.
    - *f.* Manufacture of suspender-maker's bone.
    - *g.* Manufacture of stock-maker's bone.
    - *h.* Manufacture of dress and stay maker's bone.
    - *i.* Manufacture of billiard-table cushions.
    - *k.* Manufacture of surgical instruments.
    - *l.* Manufacture of whalebone-brushes.
    - *m.* Manufacture of rosettes, woven-work, and trinkets.
    - *n.* Other whalebone manufactures.
20. PREPARATION OF TORTOISE-SHELL.

21. PREPARATION OF FISH-SCALE WORK.
22. PREPARATION OF NACRE.
23. PREPARATION OF CORAL.
24. PREPARATION OF OTHER HARD TISSUES

V. OILS AND GELATINES.

25. EXTRACTION OF WHALE-OIL, (WITH MODELS OF TRY-WORKS, CLARIFYING-VATS, &c.)
    a. Preparation of body-oil:
        Cutting in and stowing.
        Leaning and mincing.
        Trying.
        Bailing.
        Cooling.
        Barreling.
        Refining.
    b. Preparation of head-oil.
    c. Preparation of spermaceti.
    d. Instruments and appliances of rendering whale-oil:
        Boarding-knives.
        Leaning-knives.
        Mincing-horse and mincing-knives.
        Mincing-tub.
        Mincing-machine.
        Blubber-fork.
        Try-pots.
        Fire-pike.
        Stirring-pole.
        Scrap-hopper.
        Skimmer.
        Bailer.
        Cooler.
        Deck-pot.
        Casks.
26. EXTRACTION OF OTHER MAMMAL OILS.
27. EXTRACTION OF BIRD AND REPTILE OILS.
28. EXTRACTION OF FISH-OILS, (WITH MODELS OF BOILERS, PRESSES, CLARIFYING-VATS, &c.)
29. EXTRACTION OF GLUE, GELATINE, AND ISINGLASS.

VI. DRUGS, PERFUMES, AND CHEMICAL PRODUCTS.

30. MANUFACTURE OF PERFUMES.
31. MANUFACTURE OF IVORY-BLACK.
32. MANUFACTURE OF PRUSSIATES.
33. MANUFACTURE OF MUREXIDES.
34. PREPARATION OF COCHINEAL COLORS.
35. MANUFACTURE OF INKS FROM ANIMAL SUBSTANCES.
36. PREPARATION OF ALBUMEN.
37. MANUFACTURE OF PEPSIN.
38. MANUFACTURE OF PHOSPHORUS.
39. MANUFACTURE OF SAL-AMMONIAC.
40. MANUFACTURE OF AMMONIA.
41. MANUFACTURE OF ALBUMEN PREPARATIONS.
42. MANUFACTURE OF PROPYLAMINE.
43. MANUFACTURE OF FORMIC ACID.
44. MANUFACTURE OF CARBAZOTATES.

VII. MANUFACTURE OF FERTILIZERS.

45. PREPARATION OF GUANO.

a. Model of fish-guano works:

Grinders and pulverizers.

Mixers.

Guano in its various stages, with its ingredients, South Carolina phosphates, Navassa phosphates, scrap, (crude, and dried,) sulphuric acid, kainite, screened and unscreened guano, and sea-weed used in preparation.

VIII. LIMES.

46. BURNING OF LIME.

a. Models of kilns for burning shells.

IX. PRESERVATION OF THE ANIMAL FOR SCIENTIFIC USES.

47. APPARATUS FOR MAKING AND PRESERVING ALCOHOLIC SPECIMENS.

a. Tanks and jars:

Agassiz collecting-tank.

47. APPARATUS FOR MAKING AND PRESERVING ALCOHOLIC SPECIMENS—Continued.

a. Tanks and jars:

Army collecting-tank.

Museum storage-tank, Agassiz model.

Anatomical jars.

Self-sealing jars, used in collecting.

Phials.

Tube-phials.

b. Syringes for injecting.

c. Inflatable bags.

d. Preservative mixtures:

Alcohol.

Glycerine.

Carbolic acid.

Chloral hydrate.

Picric acid.

Osmic acid.

e. Labels:

Metallic labels.

Parchment labels.

Indelible inks, pencils, &c.

48. APPARATUS FOR PRESERVING AND MAKING SKELETONS.

a. Preparation of the bones:

Macerating-vats.

Boiling-vats.

Cleansing and bleaching preparation.

b. Mounting of the bones:

Scraping-tools.

Articulating-tools.

49. APPARATUS FOR MAKING CASTS. MODELING.

a. Materials:

Clays.

Plasters.

Glues.

Papier-maché and *carton pierre.*

Gelatine.

Paraffine.

Collodion.

49. APPARATUS FOR MAKING CASTS, &c.—Continued.
    b. Frames and modeling tools.
    c. Molds:
        Of plaster.
        Of gelatine.
        Of paper.
        Of paraffine.

50. APPARATUS AND METHODS OF MAKING AND MOUNTING SKINS. TAXIDERMY.
    a. Tools:
        Flaying-tools.
        Scraping-tools.
        Taxidermists' tools for stuffing:
            Forceps.
            Pliers.
    b. Preservatives and insect-powders:
        Arsenic and arsenical soap.
        Corrosive sublimate.
        Salt, alum, &c.
        Persian insect-powder.
            Syringes for application of insect-powder.
        Tobacco, snuff, used as preservatives.
    c. Frames, &c.:
        Wooden frames.
        Wire frames.
        Plaster model-bodies.

51. (ACCESSORY.) PHOTOGRAPHIC AND OTHER DELINEATING APPARATUS.
    a. Photographic apparatus:
        Lenses.
        Cameras and fittings.
        Camera tripods and stands, with model.
        Fish Commission stands.
        Plates, and their results:
            Wet plates.
            Dry plates.
            Dark closets.
    b. Camera obscuras.
    c. Mechanical delineators.
    d. Methods of heliotyping and engraving illustrations.

SECTION D.

ANIMAL PRODUCTS AND THEIR APPLICATIONS.

I. FOODS.

1. FOODS IN A FRESH CONDITION.

This section may include specimens of the marketable animals in a fresh condition in refrigerators.

For convenience in making up and arranging this portion of the collection, a list is appended of the animals used as food in the United States. Many others are available, but for various reasons are not commonly eaten.

a. Mammals:

Grizzly bear, (*Ursus horribilis.*)
Black bear, (*Ursus americanus.*)
White bear, (*Thalarctos maritimus.*)
Raccoon, (*Procyon lotor.*)
Buffalo, (*Bison americanus.*)
Musk-ox, (*Ovibos moschatus.*)
Mountain goat, (*Mazama montana.*)
Mountain sheep, (*Ovis montana.*)
Antelope, (*Antilocapra americana.*)
Moose, (*Alces malchis.*)
Woodland caribou, (*Tarandus rangifer*, subsp. *caribou.*)
Barren-ground caribou, (*Tarandus rangifer*, subsp. *grœnlandicus.*)
Elk or wapiti, (*Cervus canadensis.*)
Virginia deer, (*Cariacus virginianus.*)
Mule-deer, (*Cariacus macrotis.*)
Black-tailed deer, (*Cariacus columbianus.*)
Peccary, (*Dicotyles torquatus.*)
Manatee, (*Trichechus manatus.*)
Fox squirrel, (*Sciurus cinereus.*)
Gray squirrel, (*Sciurus carolinensis.*)

1. FOODS IN A FRESH CONDITION—Continued.

a. Mammals:

California gray squirrel, (*Sciurus fossor.*)
Tuft-eared squirrel, (*Sciurus Aberti.*)
Red squirrel, (*Sciurus hudsonius.*)
Flying squirrel, (*Sciuropterus volucella.*)
Woodchuck, (*Arctomys monax.*)
Marmots, (*Arctomys caligatus* and *flaviventer.*)
Polar hare, (*Lepus timidus*, var. *arcticus.*)
Prairie hare, (*Lepus campestris.*)
Northern hare or white rabbit, (*Lepus americanus*, and *L. americanus* var. *virginianus.*)
Red hare, (*Lepus americanus*, var. *Washingtoni.*)
Baird's hair, (*Lepus americanus*, var. *Bairdii.*)
Gray hare or gray rabbit, (*Lepus sylvaticus.*)
Sage rabbit, (*Lepus sylvaticus*, var. *Nuttalli.*)
Audubon's hare, (*Lepus sylvaticus*, var. *Auduboni.*)
Trowbridge's hare, (*Lepus Trowbridgei.*)
Jack rabbit or mule rabbit, (*Lepus callotis.*)
California hare, (*Lepus californicus.*)
Marsh hare, (*Lepus palustris.*)
Water hare, (*Lepus aquaticus.*)
Opossum, (*Didelphys virginiana.*)

b. Birds:

Reed bird or rice bird, (*Dolichonyx oryzivorus.*)
Wild pigeon, (*Ectopistes migratorius.*)
Turkey, (*Meleagris gallopavo.*)
Wild turkey, (*Meleagris gallopavo*, var. *americana.*)
Spruce grouse, (*Tetrao canadensis.*)
Dusky grouse, (*Tetrao obscurus.*)
Sage cock, (*Centrocercus urophasianus.*)
Sharp-tailed grouse, (*Pediœcetes phasianellus.*)
Prairie grouse or prairie hen, (*Cupidonia cupido.*)
Ruffed grouse, (*Bonasa umbellus.*)
Snow ptarmigan, (*Lagopus albus.*)
Rock ptarmigan, (*Lagopus rupestris.*)
White-tailed ptarmigan, (*Lagopus leucurus.*)
Bob-white or "quail," (*Ortyx virginianus.*)
Plumed partridge, (*Oreortyx pictus.*)

1. FOODS IN A FRESH CONDITION—Continued.

b. Birds:

California partridge, (*Lophortyx californicus.*)
Gambel's partridge, (*Lophortyx Gambeli.*)
Scaled partridge, (*Callipepla squamata.*)
Massena partridge, (*Cyrtonyx massena.*)
Black-billed plover, (*Squatarola helvetica.*)
Golden plover, (*Charadrius fulvus* var. *virginicus.*)
Kildeer plover, (*Aegialitis vociferus.*)
Wilson's plover, (*Aegialitis wilsonius.*)
Ringneck plover, (*Aegialitis semipalmatus.*)
Piping plover, (*Aegialitis melodus.*)
Stilt sandpiper, (*Micropalama himantopus.*)
Ruddy plover, (*Calidris arenaria.*)
Woodcock, (*Philohela minor.*)
American snipe, (*Gallinago wilsoni.*)
Red-breasted snipe, (*Macrorhamphus griseus.*)
Willet, (*Totanus semipalmatus.*)
Tell-tale, (*Totanus melanoleucus.*)
Yellow-shanks, (*Totanus flavipes.*)
Upland plover, (*Actiturus bartramius.*)
Long-billed curlew, (*Numenius longirostris.*)
Hudsonian curlew, (*Numenius hudsonicus.*)
Eskimo curlew, (*Numenius borealis.*)
Clapper rail, (*Rallus longirostris.*)
Marsh hen, (*Rallus elegans.*)
Virginia rail, (*Rallus virginianus.*)
Carolina rail, (*Porzana carolina.*)
Yellow rail, (*Porzana noveboracensis.*)
Trumpeter-swan, (*Cygnus buccinator.*)
Whistling swan, (*Cygnus americanus.*)
White-fronted goose, (*Anser albifrons.*)
Snow goose, (*Anser hyperboreus.*)
Brant, (*Branta bernicla.*)
Canada goose, (*Branta canadensis.*)
Mallard, (*Anas boschas.*)
Black duck, (*Anas obscura.*)
Pintail duck, (*Dafila acuta.*)
Gray duck, (*Chaulelasmus streperus.*)

1. FOODS IN A FRESH CONDITION—Continued.

b. Birds:

Widgeon or bald pate, (*Mareca americana.*)
Green-winged teal, (*Querquedula carolinensis.*)
Blue-winged teal, (*Querquedula discors.*)
Red-breasted teal, (*Querquedula cyanoptera.*)
Shoveller, (*Spatula clypeata.*)
Wood duck, (*Aix sponsa.*)
Big black-head, (*Fuligula marila.*)
Little black-head, (*Fuligula affinis.*)
Ring-necked duck, (*Fuligula collaris.*)
Red-head, (*Fuligula ferina*, var. *americana.*)
Canvas-back, (*Fuligula vallisneria.*)
Golden-eye, (*Bucephala clangula.*)
Barrow's golden-eye, (*Bucephala islandica.*)
Butter-ball, (*Bucephala albeola.*)
Long-tail duck, (*Harelda glacialis.*)
Harlequin-duck, (*Histrionicus torquatus.*)
Eider duck, (*Somateria mollissima.*)
King eider, (*Somateria spectabilis.*)
Scoter, (*Œdemia americana.*)
Velvet duck, (*Œdemia fusca*, var. *velvetina.*)
Surf duck, (*Œdemia perspicillata.*)
Long-billed scoter, (*Œdemia perspicillata* var. *Trowbridgei.*)
Ruddy duck or bar duck, (*Erismatura rubida.*)
Sheldrake, (*Mergus merganser.*)
Red-breasted merganser, (*Mergus serrator.*)
Hooded merganser, (*Mergus cucullatus.*)

c. Reptiles:

Gopher tortoise, (*Testudo carolina.*)
Diamond-back terrapin, (*Malacoclemmys palustris.*)
Red-bellied terrapins, (*Pseudemys rugosa.*)
Florida river-terrapin, (*Pseudemys concinna.*)
Alligator turtle, (*Macrochelys lacertina.*)
Snapping turtle, (*Chelydra serpentina.*)
Soft-shell, or leather-back turtle, (*Aspidonectes ferox*, &c.)
Green turtle, (*Chelonia mydas.*)
Pacific green turtle, (*Chelonia virgata.*)
Loggerhead turtle, (*Thalassochelys caouana.*)

1. FOODS IN A FRESH CONDITION—Continued.

d. Amphibians:

Frogs, (*Rana catesbiana, clamitans*, &c.)

e. Fishes, (eastern coast:)

File fish, (*Balistes capriscus.*)
American sole, (*Achirus lineatus.*)
Flat fish, (*Pseudopleuronectes americanus.*)
Smooth flounder, (*Pleuronectes glaber.*)
Sand flounder, (*Lophopsetta maculata.*)
Flounder, (*Chænopsetta ocellaris.*)
Southern flounder, (*Chænopsetta dentata.*)
Four-spotted flounder, (*Chænopsetta oblonga.*)
Halibut, (*Hippoglossus americanus.*)
Newfoundland "Turbot," (*Reinhardtius hippoglossoides.*)
Pollack, (*Pollachius carbonarius.*)
Cod, (*Gadus morrhua.*)
Tom-cod, or frost fish, (*Microgadus tomcodus.*)
Haddock, (*Melanogrammus æglefinus.*)
Hake, (*Phycis chuss.*)
Squirrel hake, (*Phycis tenuis.*)
Cusk, (*Brosmius americanus.*)
Whiting, (*Merlucius bilinearis.*)
Norway haddock, (*Sebastes norvegicus.*)
Rose fish, (*Sebastes viviparus.*)
Tautog, or black-fish, (*Tautoga onitis.*)
Chogset, or cunner, (*Tautogolabrus adspersus.*)
Hog fish, (*Lachnolæmus falcatus.*)
Angel fish, (*Holacanthus ciliaris.*)
Sword fish, (*Xiphias gladius.*)
Spear fish, (*Tetrapturus albidus.*)
Sail fish, (*Histiophorus americanus.*)
Mackerel, (*Scomber scombrus.*)
Chub mackerel, (*Scomber colias.*)
Bonito, (*Sarda pelamys.*)
Horse mackerel, (*Orcynus secundi-dorsalis.*)
Spanish mackerel, (*Cybium maculatum.*)
Cero, (*Cybium caballa.*)
Striped cero, (*Cybium regale.*)
Crevallé. (*Carangus hippos* and *Paratractus pisquetus.*)

1. FOODS IN A FRESH CONDITION—Continued.

e. Fishes, (eastern coast:)

Pompano, (*Trachynotus carolinus.*)
Short pompano, (*Trachynotus ovatus.*)
Butter-fish, (*Poronotus triacanthus.*)
Squeteague, (*Cynoscion regalis.*)
Spotted squeteague, (*Cynoscion carolinensis.*)
Drum, (*Pogonias chromis.*)
Spot, (*Liostomus obliquus.*)
Silver perch, or yellow-tail, (*Bairdiella punctata.*)
Red fish, or spotted bass, (*Sciænops ocellatus.*)
King fish, (*Menticirrus nebulosus.*)
Southern king fish, or Bermuda whiting, (*Menticirrus alburnus.*)
Croaker, (*Micropogon undulatus.*)
Sailor's choice, (*Lagodon rhomboides.*)
Sheeps-head, (*Archosargus probatocephalus.*)
Scuppaug, or porgy, (*Stenotomus argyrops.*)
Grunts, (*Hæmulon arcuatum*, &c.)
Gray snapper, (*Lutjanus caxis.*)
Red snapper, (*Lutjanus aya.*)
Grouper, (*Epinephelus morio.*)
Spotted grouper, (*Epinephelus guttatus.*)
Jew fish, (*Promicrops guasa.*)
Sea bass, (*Centropristis atrarius.*)
Squirrel, (*Diplectrum fasciculare.*)
Striped bass or rock fish, (*Roccus lineatus.*)
White perch, (*Morone americana.*)
Moon fish, (*Parephippus quadratus* and *P. faber.*)
Triple-tail, (*Lobotes surinamensis.*)
Blue fish, (*Pomatomus saltatrix.*)
Striped mullet, (*Mugil lineatus.*)
Silver-sides, (*Chirostoma notatum.*)
Silver gar fish, (*Belone longirostris.*)
Skipper, (*Scomberesox scutellatus.*)
Mummichogs, (*Hydrargyra majalis*, &c.)
Capelin, (*Mallotus villosus.*)
Smelt, (*Osmerus mordax.*)
Salmon, (*Salmo salar.*)

1. FOODS IN A FRESH CONDITION—Continued.

e. Fish, (eastern coast:)

Sea trout, (*Salmo immaculatus.*)
Tarpum, (*Megalops thrissoides.*)
Menhaden, (*Brevoortia menhaden.*)
Shad, (*Alosa sapidissima.*)
Alewife, or gaspereau, (*Pomolobus pseudoharengus.*)
Tailor herring, (*Pomolobus mediocris.*)
Herring, (*Clupea harengus*)
Mud shad, (*Dorosoma cepedianum.*)
Anchovy, (*Engraulis vittatus*, &c.)
Sea eel or conger, (*Conger oceanica.*)
Eel, (*Anguilla bostoniensis.*)
Sturgeon, (*Acipenser oxyrhynchus* and *A. brevirostris.*)
Lamprey eel, (*Petromyzon americanus.*)

f. Fishes, (fresh waters:)

Burbot or lawyer, (*Lota maculosa.*)
Fresh-water drum, (*Haploidonotus grunniens.*)
Small-mouthed black-bass, (*Micropterus salmoides.*)
Large-mouthed black-bass, (*Micropterus floridanus.*)
Rock-bass, (*Ambloplites rupestris.*)
Sacramento "perch," (*Archoplites interruptus.*)
Sun-fish, (*Pomotis aureus.*)
Black-eared sunfish, (*Pomotis auritus.*)
"Bream" of Southern States, (*Calliurus*, *Lepomis*, *Enneacanthus*, *Chænobryttus*, numerous species.)
Strawberry or grass bass, (*Hyperistius hexacanthus*, and *Pomoxys storerius.*)
Yellow perch, (*Perca flavescens.*)
Yellow pike-perch, (*Stizostedium americanum.*)
Gray pike-perch or sauger, (*Stizostedium griseum.*)
Canada pike-perch, (*Stizostedium canadense.*)
White bass, (*Roccus chrysops.*)
Short-striped bass, (*Morone interrupta.*)
Lake pike, (*Esox lucius.*)
Pickerel, (*Esox reticulatus*, *E. fasciatus*, *E. cypho*, &c., &c.)
Masquallonge, (*Esox nobilior.*)
Brook trout, (of eastern slope,) (*Salmo fontinalis.*)
Brook trout, (of western slope,) (*Salmo iridea.*)

1. FOODS IN A FRESH CONDITION—Continued.

f. Fish, (fresh waters:)

Utah trout, (*Salmo virginalis.*)

Oquassa trout, (*Salmo oquassa.*)

Lake trout, (*Salmo confinis.*)

Salmon trout or Mackinaw trout, (*Salmo namaycush.*)

Siscowet, (*Salmo siscowet.*)

Sebago salmon, (*Salmo sebago.*)

Missouri trout, (*Salmo Lewisi.*)

White fish, (*Coregonus albus.*)

Otsego white fish, (*Coregonus otsego.*)

Lake herring, (*Argyrosomus harengus* and *A. clupeiformis.*)

Black fin of Lake Michigan, (*Argyrosomus nigripinnis.*)

Michigan grayling, (*Thymallus tricolor.*)

Mountain grayling, (*Thymallus montanus.*)

Suckers of eastern slope, (*Catostomus teres*, &c., *Ptychostomus aureolus*, &c.)

Suckers of western slope, (*Catostomus occidentalis*, &c.)

Fall fish, (*Semotilus rhotheus.*)

Chubs of eastern slope, (*Semotilus corporalis*, &c.)

Chubs of western slope, (*Lavinia exilicauda, Algansea*, sp., &c.)

"Pike" or "salmon trout" of California, (*Ptychocheilus grandis*, &c., *Pogonichthys inæquilobus*, &c.)

Dace, (*Ceratichthys biguttatus*, &c.)

Buffalo fish, (*Bubalichthys bubalus.*)

Shiner, (*Stilbe americana.*)

Carp, (*Carpiodes cyprinus*, &c.)

Catfishes, (*Amiurus catus*, *A. nigricans*, &c., *Ictalurus cœrulesceus*, &c., and many other siluroid fishes.)

Sturgeon of the lakes, (*Acipenser rubicundus.*)

Shovel-nose sturgeon, (*Scaphirhynchops platyrhynchus.*)

g. Fishes, (western coast:)

Flounders, (*Platichthys stellatus*, *Lepidopsetta umbrosa*, &c.)

"Soles," (*Parophrys vetulus*, *Psettichthys melanostictus*, &c.)

Halibut, (*Uropsetta californiana*, *Hippoglossus*, sp., &c.)

Tomcod, (*Microgadus proximus.*)

Cod of Alaska, (*Gadus macrocephalus.*)

Rock fish or "rock cod," (*Sebastomus rosaceus* and species of *Sebastosomus*, *Sebastichthys*, &c.)

1. FOODS IN A FRESH CONDITION—Continued.

g. Fishes, (western coast:)

Rock trout, (*Chirus constellatus.*)

"Cod" of San Francisco, (*Ophiodon elongatus.*)

Black fish or "sheeps-head," (*Pimelometopon pulcher.*)

"Perch," (numerous species of *Embiotoca*, *Holconotus*, &c.)

"Bass," (*Atractoscion nobilis.*)

Cognard or little bass, (*Genyonemus lineatus.*)

San Francisco "smelt," (*Atherinopsis californiensis.*)

Pacific smelt, (*Osmerus elongatus.*)

Salmon, (*Salmo quinnat*, &c.)

Oulachan, (*Thaleichthys pacificus.*)

Sardine or pilchard, (*Pomolobus cœruleus.*)

Herring, (*Clupea mirabilis.*)

Sturgeon, (*Acipenser acutirostris*, &c.)

Columbia River sturgeon, (*Acipenser transmontanus.*)

h. Crustaceans.[1]

i. Mollusks.[1]

2. FOODS: DRIED AND SMOKED.

a. Mammal preparations:

Jerked bear-meat.

Jerked seal and walrus meat, (Indian.)

Jerked and smoked buffalo-meat.

Dried and smoked beef.

Dried and smoked venison.

Hams of various kinds.

Jerked porpoise-meat, (Indian.)

Jerked squirrels and other small mammals.

Pemmican.

Meat-biscuit, desiccated meat, meat extract, (*extractum carnis*,) desiccated milk, &c.

Sausages.

Cheese.

b. Bird preparations:

Jerked birds, (Indian.)

---

[1] The various applications of these groups are enumerated in the "*List intended to give a general idea of the useful products (other than vertebrates) of the sea and shore, as well as of the interior waters of the United States*," prepared by Mr. WM. H. DALL, and printed as Circular No. 2 of series (C,) National Museum series.

2. FOODS: DRIED AND SMOKED—Continued.

c. Reptile preparations:

Dried lizards, (Indian.)

d. Fish preparations:

*Smoked halibut.

Dried cod, haddock, hake, &c.

Dried and smoked mullet and roes.

Dried and smoked garfish, flying-fish, &c.

Smoked herring, alewives, &c., and their roes.

Smoked salmon, oulachan, white-fish, smelt, &c., and their roes.

Smoked sturgeon.

Veziga, prepared from the notochord of sturgeons.

e. Insects:

Dried grasshoppers, (Indian.)

f. Worms:

Dried worms, (Indian.)

g. Mollusk preparations:

Dried abalones, (*Haliotis*,) prepared by the California Chinese.

Dried siphons of *Schizothærus* prepared by the Indians of the northwest coast.

Dried slugs, (*Limax*, &c.,) used by Indians.

h. Radiate preparations:

(Dried holothurians, "bêches de mer," used by Chinese.)

i. Protozoans:

("Mountain meal," a kind of infusorial earth, mixed with flour, and used as food in Lapland and China.)

3. FOODS: SALTED, CANNED, AND PICKLED.

a. Mammal preparations:

Salted buffalo-meat.

Salted beef.

Salted deer, reindeer, elk.

Salted tongues of beef, buffalo, deer, horse.

Salted pork.

Canned milk of the various brands.

b. Bird preparations:

Canned turkey.

Canned chicken.

Canned goose.

3. FOODS: SALTED, CANNED, AND PICKLED—Continued.

b. Bird preparations:

(Canned ortolans, (*Emberiza hortularia,*) esteemed a delicacy in Cyprus.)

c. Reptile preparations:

Salted and canned turtles and turtle soup.

Canned frogs.

d. Fish preparations:

Salted halibut, halibuts' fins, &c.

Salted cod, cods' tongues, sounds, and roe.

Salted mackerel.

Salted Spanish mackerel.

Salted bluefish.

Salted pompano.

Salted sword-fish.

Salted mullets.

Salted salmon.

Salted white-fish.

Salted trout.

Salted shad.

Salted herring.

Salted gaspereau.

Salted menhaden.

Salted anchovies.

(Spiced lampreys) used in Europe.

Anchovy-sauce and "essence of anchovies."

Canned menhaden, in oil, "American sardines."

Canned menhaden, in oil, "American club-fish."

Spiced menhaden, "ocean trout."

Canned herring, in oil, "Russian sardines."

Caviare, prepared from roe of the various sturgeons.

(Caviare, prepared from roe carps, used by Jews.)

("Boutargue" or "botargo" prepared on the Mediterranean from the roes of *Labrax* and *Mugil.*)

e. Crustacean preparations:

Canned lobsters.

Canned crabs.

Canned prawns and shrimps.

f. Mollusk preparations:

Canned oysters.

3. FOODS: SALTED, CANNED, AND PICKLED—Continued.

f. Mollusk preparations:

Canned clams.

Canned little-neck clams.

Canned scollops.

(Cockles, (*Cardium edule,*) used in Europe as pickles and catsup.)

4. GELATINES.

a Mammal gelatines, (see also under 24:)

Gelatines made from tanners refuse and from sinews.

Gelatines made from feet and hoofs.

Gelatines made from bone and ivory shavings.

b. Bird gelatines:

(Nests of esculent swallows, (*Calocalia esculenta, C. fuciphaga, C. indifica,* &c.,) exported from Indian Archipelago to China.)

c. Fish gelatines or isinglass, (see also under 24.)

d. Insect gelatine:

Gelatine from cocoons of silk-worms.

5. BAITS AND FOODS FOR ANIMALS.

a. Prepared baits, (see under B, 45.)

b. Food for domesticated animals:

Oil-factory scraps.

Fish-scraps.

Cuttle-fish bone, (see under 18.)

II. CLOTHING.

6. FURS, (embracing the furs in their rough state, (*peltries,*) and in the various stages of preparation; also the manufactured articles, such as robes, rugs, cloaks, sacks, tippets, cuffs, muffs, hats, caps, gloves, trimmings and linings.)[1]

a. Mammal furs:

(Diana monkey, (*Cercopithecus diana,*) of West Africa.)

(Black monkey, (*Colobus polycomus,* and other species,) of West Africa—trimmings, &c.)

(Abyssinian monkey, (*Colobus guereza.*))

---

[1] *Note.*—For convenience in arranging the general collections of the museum, this list has been made unusually full, and includes all furs known to be found in American and European markets.

6. FURS—Continued.

a. Mammal furs:

(American howling-monkey, (*Mycetes*, several species)—muffs.)

(Lion, (*Felis leo*,) of Africa and Asia—rugs.)

(Tiger, (*Felis tigris*)—rugs, &c.)

(Leopard, (*Felis pardus*)—rugs and saddle-cloths.)

Puma, (*Felis concolor*)—carriage-robes, rugs, &c.

Ocelot, (*Felis pardalis*)—rugs.

Jaguar, (*Felis onca*)—rugs.

Cat, (*Felis domestica*)—robes and philosophical apparatus.

Black cat.

White cat.

Maltese cat.

Tortoise-shell.

(Wild-cat, (*Felis catus*,) of Europe and Asia—robes and linings.)

(Snow leopard, (*Felis irbis*,) of Asia.)

Eyra, (*Felis eyra*.)

Yaguarundi, (*Felis yaguarundi*.)

(Cheetah, (*Cynailurus jubatus*,) of India and Southern Asia.)

Bay lynx, (*Lynx rufus*)—rugs, and, when dyed, muffs and boas.

Canada lynx, (*Lynx canadensis*)—rugs and trimmings, and dyed muffs, boas, &c.

Dog, (*Canis familiaris*.)

Eskimo dog.

Wolf, (*Canis lupus*)—linings, rugs, and robes.

White wolf.

Black wolf.

Gray wolf.

"Blue wolf."

Red wolf.

Coyote, or prairie wolf, (*Canis latrans*)—rugs and robes.

(Jackal, (*Canis aureus*,) of Old World.)

Red fox, (*Vulpes alopex*, var. *fulvus*)—robes, (mostly imported to Turkey.)

Cross fox, (*Vulpes alopex*, var. *decussatus*)—robes, trimmings.

Black and silver fox, (*Vulpes alopex*, var. *argentatus*)—muffs, cloaks, trimmings; also, fox-skins dyed to imitate lynx; also, various imitations of silver-fox, made from skins of more common varieties.

6. FURS—Continued.

a. Mammal furs:

Arctic fox, (*Vulpes lagopus.*)

White fox.

Blue fox.

Kit fox, (*Vulpes velox*)—robes, muffs, trimmings.

(Cossac fox, (*Vulpes corsac*,) of Asia.)

(Mountain fox, (*Vulpes montanus*,) of India.)

Gray fox, (*Urocyon virginianus*)—rugs, robes, and linings.

(Spotted hyena, (*Hyæna crocuta*,) of West and South Africa.)

(Striped hyena, (*Hyæna striata*,) of West Africa and India.)

Fisher or pekan, (*Mustela Pennanti*)—linings, tails used for trimmings.

American or Hudson's Bay sable, (*Mustela americana*)—cloaks, muffs, cuffs, boas, linings, &c.:

Silver variety.

Orange variety.

Brown or common variety.

(Russian sable, (*Mustela zibellina*,) of North Europe and Asia—cloaks, muffs, boas, linings, &c.)

(Tartar sable, or kolinsky, (*Mustela sibirica*)—cloaks, muffs, and dyed to imitate Russian sable.)

(Pine marten, (*Mustela abietum*,) of North Europe and Asia.)

(Stone marten, or French sable, (*Mustela saxorum*,) of Europe—dyed to imitate sable.)

(Beech marten, (*Mustela foina*,) of Europe and Asia—dyed to imitate sable.)

(Polecat, fitch, or ferret, (*Putorius vulgaris*,) of Europe and Asia.)

Ermine, or weasel, (*Putorius erminea*,) of Northern Hemisphere—cloaks, linings, &c.:

Royal ermine, trimmed with astrakhan fur, (miniver.)

Siberian ermine.

Long-tailed weasel, (*Putorius longicauda*:)

Summer dress.

Winter dress.

Mink, (*Putorius vison*,)—cloaks, muffs.

Wolverine, (*Gulo luscus*,)—muffs, robes, linings.

American badger, (*Taxidea americana*)—muffs and rugs.

6. FURS—Continued.

a. Mammal furs:

(European badger, (*Meles vulgaris*)—muffs and rugs.)

Skunk, Alaska sable, (*Mephitis mephitica*)—muffs, boas, &c.

White-backed skunk, (*Conepatus mapurito.*)

Striped skunk, (*Spilogale putorius.*)

Otter, (*Lutra canadensis,*) with specimens of the plucked and dyed fur—muffs, trimmings, &c.

Sea otter, (*Enhydra marina*)—muffs, gloves, collars, cuffs, trimmings.

Black bear, (*Ursus americanus*)—caps, rugs, muffs, robes, &c.

a'. Cinnamon variety.

b. Silvery variety.

(Brown bear, (*Ursus arctos,*) of Europe and Asia.)

Grizzly bear, (*Ursus horribilis*)—rugs, robes, trimmings.

White bear, (*Thalarctos maritimus*)—rugs, robes, and used extensively by the Eskimos.

Raccoon, (*Procyon lotor*)—hats, linings.

Fur-seal, (*Callorhinus ursinus*)—cloaks, hats, gloves, muffs, linings, trimmings, &c.

Cub fur.

(Antarctic fur-seal, (*Arctocephalus aucklandicus*, &c.)

Hair seal, (*Phoca vitulina* and *Phoca Richardsii*)—coats, caps, linings for shoes.

Harp seal, (*Pagophilus grœnlandicus,*) with specimens of the white fur of the unborn cub, and the blue fur of the young.

Hood seal, or bladder-nose, (*Cystophora cristata.*)

Square flipper, or bearded seal, (*Erignathus barbatus,*) with specimens of fur dyed to imitate leopard.

Banded seal, (*Histriophoca equestris*)—used by Eskimos as fur.

Gray seal, (*Pusa gryphus.*)

Ringed seal, (*Pagomys fœtidus.*)

Bison, or buffalo, (*Bison americanus*)—rugs and robes.

a'. Mountain bison.

b. Common bison.

Musk-ox, (*Ovibos moschatus*)—robes, rugs, and trimmings.

(Yak, (*Poëphagus grunniens,*) of Asia—robes and trimmings.)

Mountain-goat, (*Aplocerus montanus*)—robes, &c.

6. FURS—Continued.

a. Mammal furs:

(Llama, guanaco, paco, and vicugna, (*Auchenia*, sp.)—trimmings, &c.)

Goat, (*Capra*, sp.)—rugs, trimmings.

a'. Angora goat.

b. Cashmere goat.

c. Other varieties.

Sheep, (*Ovis aries*)—rugs, trimmings, &c.

a. Astrakhan sheep.

b. Caracoul sheep.

c. Other varieties. Lamb-skins and dyed furs.

Antelope, (*Antilocapra americana*)—rugs.

Moose, (*Alces malchis*)—rugs and robes.

Elk, (*Cervus canadensis*)—rugs and robes.

Reindeer, (*Tarandus rangifer*)—robes, coats, gloves, &c.

Caribou, (*Tarandus rangifer* var.)—robes, coats, gloves.

Mule deer, (*Cariacus macrotis*)—trimmings, robes.

Virginia deer (*Cariacus virginianus*)—trimmings, robes.

Mole, (*Scalops* and *Condylura*, sp.)—robes, garments.

(European mole, (*Talpa europæa*)—robes, garments.)

Woodchuck, or siffleur, (*Arctomys monax*)—robes, exported to Europe as "white and gray weenusk."

Marmot, (*Arctomys caligatus*)—robes, trimmings.

Parry's marmot, (*Spermophilus Parryi*)—robes, trimmings.

Gray squirrel, (*Sciurus carolinensis*, &c.)—trimming; tails used for boas.

(Squirrel, or "calabar," (*Sciurus vulgaris*,) Northern Europe and Asia.)

a'. Siberian squirrel. Trimmings, muffs, capes, &c.; tails used for boas, dyed to imitate sable.

b. "Weisenfels linings" of the white fur of the belly.

Showt'l, (*Haplodontia leporina*)—used by Indians.

(Chinchilla, (*Chinchilla laniger*,) of South America—muffs, mantles, boas, cloak-linings, and trimmings.)

Musquash, (*Fiber zibethicus*)—muffs, capes, caps, and linings, and imitations of beaver-fur.

(Neutria, or Coypu, (*Myopotamus coypus*)—linings and muffs, and imitations of beaver.)

6. FURS—Continued.

a. Mammal furs:

(Beaver, (*Castor fiber*,) of Northern Europe and Asia.)

Beaver, (*Castor canadensis*)—linings and muffs.

White beaver.

Spotted beaver.

Rats and mice, (*Mus.*, sp. var.)

Lemming, (*Myodes torquatus* and *obensis*)—robes.

Rabbit, or cony, (*Lepus cuniculus*)—children's furs, and imitations of seal, beaver, &c., exported largely to China.

White variety.

Blue variety.

Brown variety.

American native rabbit furs, such as *Lepus glacialis*, used for muffs, boas, and feltings.

Possum, (*Didelphys virginiana.*)

(Kangaroo, (*Macropus giganteus*,) of Australia.)

(Ornithorhynchus, (*Ornithorhynchus anatinus*,) of Australia.)

b. Skins of birds used as furs:

Turkey furs, (*Melagris gallopavo*, &c.)

Gull furs, (*Larus argentatus*, &c.)

Grebe furs, (*Podiceps aristatus*, &c.)

Loon furs, (*Colymbus torquatus*, &c.)

Swan furs and swan's down trimmings, (*Cygnus americanus*, &c.)

Pelican furs, (*Pelecanus fuscus*, &c.)

Adjutant crane, (*Ciconia argala*)—feathers used as fur.

Puffin furs, (*Fratercula arctica*, &c.)

Penguin furs, (*Aptenodytes*, *Pennantii*, &c.)

Feathers of common fowl used in trimmings.

7. LEATHERS. (See under 20.)

8. TEXTILE FABRICS.

a. Prepared from hair of mammals:

Human hair used in manufacture of watch-chains.

Hair of bats used in felting and in plaiting ropes in Central America and tassels in New Caledonia.

Hair of raccoon used in felting, (largely exported to Germany for the use of hatters.)

Hair of weasels and sables used in felting.

8. TEXTILE FABRICS—Continued.

a. Preparations of hair of mammals:

Hair of fur seal woven with silk in the manufacture of shawls.

Moose hair and its fabrics.

Ox and calf hair used in the manufacture of imitation woolen goods.

Sheep's wool, with specimens of fleeces and stapled wools, from various breeds and localities, short-wool fabrics, broadcloths, merinoes, flannels, mouselins de laine, serges, tweeds, blankets, carpets, and tartans, worsted fabrics, stuffs, bombazines, camlets, shawls, plushes and velvets, hosiery, and yarns, felts, felt-cloths, and felt-hats.

Goats' wool with specimens of mohairs, cashmeres, plushes, velveteens, camlets, and shawls. (For manufactured wigs and perukes, see under 21.)

(Yak (*Poëphagus grunniens*) wool with specimens of yak-lace and other fabrics.)

(Camels' hair with specimens of fabrics, plushes, felts, shawls, &c.)

(Hair of llama, paco, guanaco, and vicugna, with specimens of alpaca, guanaco, and other fabrics, and umbrellas and other articles manufactured.)

Hair of horses used in weaving furniture-covers, crinoline-skirts, and bags for pressing oil.

Hair of buffalo used in plaiting ropes, lariats, &c.

Fur of mole used in felting.

Beaver (castor) fur with specimens of the felt cloths, hats, &c.

(Neutria-fur used in felting and in the manufacture of hats.)

Musquash fur used in felting.

Possum hair with fabrics of Indian and other manufacture.

Fur of rabbit and hare used in felting, with specimens of hats and cloths.

Whalebone fiber used in weaving cloth covers for telescopes, &c.

b. Prepared from feathers of birds:

Cloths woven from feathers, (China.)

c. Prepared from silk of insects: (This collection should include specimens of the cocoons, the raw silk, the spun silk, and of the various fabrics, plain and figured silks, satins and satinettes, shawls, damasks, brocades, crapes, and ribbons.)

8. TEXTILE FABRICS—Continued.

c. Prepared from silk of insects:

Silk of common silk-worm, (*Bombyx mori.*)

Silk of *Samia cecropia*, *Samia polyphemus*, and other native American moths.

(Silk of exotic moths other than *Bombyx mori*, such as the tussah, (*Bombyx pernyi* and *Bombyx mylitta*,) the moonga, (*Saturnia assamensis*,) the joree, (*Bombyx religiosa*,) the ena or arindy, (*Bombyx cynthia*.))

Fabrics woven by the insects themselves, as *Tinea padilla.*

Silk of spiders.

d. Prepared from byssus of mollusks.

(Fabrics woven from byssus of the wing-shell (*Pinna nobilis*) and other mollusks.)

III. MATERIALS EMPLOYED IN THE ARTS AND MANUFACTURES.

* *Hard materials.*

9. IVORY AND BONE. (This collection should include specimens of the various ivories and bones in their rough state, and manufactured into buttons, trinkets, cutlery-handles, canes, pen and pencil handles, brush-handles, billiard and bagatelle balls, dice, piano-keys, harness-rings, combs, false-teeth, philosophical instruments, and as used by portrait painters and photographers.)

a. Ivory of mammals:

Tusks of walrus used for trinkets, handles, jewelry, buttons, paper-knives, counters, &c.

Teeth of bears, dogs, wolves, foxes, peccaries, and other large mammals, used as implements, arrow-tips, and ornaments, by Indians.

Elk-ivory used by Indians in ornamentation.

Tusks of mammoth elephant (*Elephas primigenius*) from northern America and Asia, with Eskimo carvings and specimens of "Siberian ivory."

9. IVORY AND BONE—Continued.

a. Ivory of mammals:

(Tusks of African elephant with specimens of sawed and scroll ivory and of the manufactured balls, combs, piano-keys, handles, rings, canes, buttons, trinkets, bangles, and miniature tablets.)

(Tusks of the Asiatic elephant and their applications.)

(Teeth of hippopotamus as used for handles for surgical instruments, index-fingers, and formerly for false-teeth, (trade-name, "sea-horse.")

Teeth of wild-hog used in manufacture of jewelry, vinaigrettes, &c.

Teeth of peccary.

Ivory of narwhal used for canes.

Teeth of sperm-whale and their application to the manufacture of balls, buttons, and trinkets.

Incisors of beaver used by Indians for chisels, knives, and ornaments.

b. Ivory of reptiles:

Teeth of alligator used for jewelry, whistles, cane-handles, buttons, &c.

c. Ivory of fishes:

Sharks' teeth used in arming weapons.

Teeth of sharks and other fish used as trinkets.

Jaws of the sleeper-shark (*Somniosus brevipinna*) used for head-dresses by Indians.

d. Bone of mammals:

Parts of splanchno-skeleton of ferae, used as charms.

Bones of bear and other large mammals, used by Indians for implements, and as tablets for paintings.

Bones of buffalo and of the domestic ruminants, used as substitute for ivory in the manufacture of buttons, handles, combs, &c.

Sperm-whale jaw-bone, used for harness-rings, martingales, &c.

Horn-cores of ruminants, used in manufacture of assayers' cupels.

e. Bone of birds:

Bones of birds, used by Indians and Eskimos in making awls, needles, flutes, bird-calls, and dress-trimmings.

9. IVORY AND BONE—Continued.

f. Bone of fishes:

Fish-bones, used by Indians and Eskimo in making implements.

Shark's vertebræ, used for canes.

Bones of sharks and skates, used (in Japan) in making imitation tortoise-shell.

g. Waste bone and ivory:

Use in manufacture of bone-black, ivory-black, and bank-note ink, (see under 29.)

Use in manufacture of sizes and glues, (see under 24.)

Use in manufacture or gelatine for food, (see under 4.)

Use in manufacture of phosphorus, carbonate of ammonia, (hartshorn,) and sal ammoniac, (see under 30.)

Use in manufacture of bone-charcoal for filters, (see under 30.)

Use in manufacture of paper.

Use of shavings in case-hardening gun-barrels and other fine steel.

10. HORN. (Embracing the varieties of horn known to commerce, the split and pressed horns, and the various manufactured articles, such as jewelry, combs, and handles.

a. Horn, employed as a material:

Horn of rhinoceros, used for handles and trinkets, cups, boxes, whips, and canes.

Horns of ox, sheep, and goat, used for handles, buttons, combs, powder-flasks, cups, boxes, stirrups, spoons, and imitations of tortoise-shell, also "sensitive Chinese leaves," and formerly for transparent plates in lanterns and horn-hooks, for trumpets, and for finger-nails in lay figures.

Horn of buffalo, used like that of ox.

(Horn of Asiatic buffalo, (*Bos bubalus.*))

Horn of mountain-sheep and mountain-goat, used by Aleutians, in making spoons, bowls, and numerous other implements.

b. Antlers:

Antlers of deer, elk, and moose, (stag-horn,) used in the manufacture of handles for instruments, trinkets, and buttons.

10. HORN—Continued.

b. Antlers:

Antlers of deer, elk, moose, and nearly all species of ruminants, employed for ornamental purposes.

c. Chemical and other applications:

Burnt horn, (*cornu ustum*,) used in dentifrices.

Carbonate of ammonia, (hartshorn,) manufactured from deer-horns, (see under 30.)

11. HOOFS AND CLAWS, &c. (Embracing the commercial hoof, and the various stages of manufacture represented by specimens.)

a. Hoofs:

Hoofs of ox and bison, used in making buttons, combs, and handles.

Hoofs of horse, used like those of ox and bison.

Hoofs of musk-ox, deer, and antelope, used by Indians in ornamentation.

Feet of deer, used for knife-handles, stool-feet, &c.

b. Claws:

Claws of bear, puma, wolf, &c., used by Indians in ornamentation.

(Claws of lion and tiger, used by jewelers for trinkets.)

Human nails, used by Indians for ornamental trimmings.

c. Chemical applications of hoofs and claws:

Use in manufacture of prussiate of potash, (see under 30.)

Use in manufacture of glue, (see under 24.)

12. BALEEN. (Embracing the commercial baleen in its various grades, *Greenland*, *Northwest Coast*, *South Sea*, *fin-back*, and *hump-back*, with the split, twisted, and dyed bone.)

a. Whalebone, as used by manufacturers of ribbons, hats, umbrellas, whips, canes, boots, fishing-rods, billiard-tables, buttons, handles, brushes, surgical instruments, stays, corsets, crinolines, harness-rosettes, covers, stuffings, light woven hats and bonnets, &c.; also, imitation whalebone, (wallosin,) made from rattan.

13. TORTOISE-SHELL. (Embracing the carapace entire, and the commercial shell, *blades*, *feet*, *noses*, and *head*.)

13. TORTOISE-SHELL—Continued.

a. Shell of tortoise (*Eretmochelys imbricata, E. squamata*) used in manufacture of combs, handles, jewelry, inlaying, and buttons, together with imitations of tortoise-shell in horn, shark's bone, and celluloid.

b. Shells of land tortoises, used by Indians for pots, scoops, and rattles.

14. SCALES.

a. Shell of mammals:

Shell of armadillo, used by Texans and Mexicans.

b. Scales of fishes used in ornamental work, with specimens of flowers and other articles manufactured:

Scales of parrot fishes, (*Scaridæ* and *Labridæ.*)

Scales of mullets, (*Mugilidæ.*)

Scales of sheepshead, &c., (*Sparidæ.*)

Scales of drum and bass, (*Sciaenidæ.*)

Scales of Serranidæ and perches, (*Percidae* and *Labracidæ.*)

Scales of Lobotidæ.

Scales of tarpum, (*Elopidæ.*)

Scales of herrings, (*Clupeidæ.*)

Scales of Cyprinidæ.

Scales of eels, used in the north of Europe to give a pearly luster in ornamental house-painting.

Scales of gar-pikes, used by Indians for arrow-tips.

(Pearl white, or *essence d'Orient*, prepared from scales of *Alburnus lucidus* and other Cyprinidæ and Clupeidæ, used in making artificial pearls.) (See under 27.)

Shagreen of trigger-fish, (*Balistes*,) used in polishing wood.

Shagreen of sharks, used as leather, (see under II, B. 5,) and for polishing purposes, particularly in the manufacture of quill pens.

Scales of sturgeons, used by Indians for implements.

For gelatine as a material and the arts and papier glacé, see 24.

15. PEARL.

a. Pearls and nacre, (embracing the pearl-yielding shells, with the pearls and the mother-o'-pearl in the rough state, with the manufactured buttons, handles, and jewelry, pearl-powder, inlaid work, and papier maché, ornamented with mother-o'-pearl:)

15. PEARL—Continued.

a. Pearls and nacre:

Top-shells, (*Turbinidæ,*) and their application to manufacture of shell-flowers.

Tower-shells, (*Trochidæ.*)

Ear-shells, (*Haliotidæ,*) used in manufacture of buttons, handles, inlaid work, and pearl-powder.

Other gastropods supplying nacre.

Pearl-oysters, (*Aviculidæ,*) with pearls and nacre.

River-mussels, (*Unionidæ,*) with pearls and nacre.

Mussels, oysters, and other conchifers supplying pearls and nacre.

Shells of nautilus and argonaut, prepared to exhibit their nacre.

Ornamental pearl-work, imitating sprays of flowers, &c.

Imitation pearls.

16. SHELL.

a. Cameo shell:

Shell of conch, (*Strombus gigas,*) and carvings.

Shell of helmet, (*Cassis rufa*, *C. tuberosa*, and *C. madagascariensis,*) with carvings.

b. Shells used for implements, &c.:

Shells of *Strombus*, *Triton*, *Dolium*, *Fusus*, *Murex*, and *Buccinum*, used for fog-horns, lamps, vases, and ornamental borders in flower-gardens.

Shells of *Busycon*, *Sycotypus*, *Mactra*, &c., used by Indians in manufacture of implements, with specimens of implements.

Shells of *Mactra*, used for ladles, scoops, and spoons by fishermen.

Shells of *Tridacna*, used for vases, fountains, and in the manufacture of handles and carvings.

Shells of *Pecten*, *Haliotis*, *Dentalium*, *Mercenaria*, &c., used by Indians for trimmings and ornaments.

(Scallop, or palmer's shell, (*Pecten jacobæus,*) used as a decoration of honor.)

(Chank shell, (*Turbinella pyrum,*) used in the manufacture of Hindoo bangles, and in polishing cloth.)

Shells of *Pecten*, used in making pin-cushions and purses.

16. SHELL—Continued.

b. Shells used for implements, &c.:

(Painters' mussel, (*Unio pictorum*,) used to hold colors.)

(Shells of *Placuna placenta*, used in China as a substitute for window-glass.)

Shells of *Mercenaria violacea*, *Purpura lapillus*, and *Buccinum undatum*, used by Indians of eastern coast in manufacture of money, with specimens of wampum, (with the modern wampum or shell-beads manufactured for the Indian trade,) and of the hyqua or *Dentalium* shells, employed in a similar manner by the Indians of the Pacific coast.

Specimens of the cowry, (*Cypraca moneta.*) "Live cowry" and dead cowry, used in African trade and for trimmings.

Shells of *Cypraea*, *Rotella*, *Oliva*, *Turritella*, *Phasianella*, (Venetian shells,) &c., mounted as buttons and jewelry.

Composition shell-work for box-covers and frames, made by glueing shells in mosaics.

Calcined shells, used by dentifrice and porcelain makers. (See, also, under 32.)

Cuttle-fish bone from *Sepia officinalis*, used as a pounce, as a dentifrice, as polishing-powders, for taking fine impressions in counterfeiting, and as food for birds. (See, also, under D. 5.)

Concretions from the stomach of *Astacus*, known as "crab's-eyes" and "crab-stones," and used as antacids.

Shell of king-crab, (*Limulus polyphemus*,) used as a boat-bailer.

Opercula of mollusks, used as "eye-stones."

17. CORAL.

a. Coral as a material:

Red coral, (*Corallium nobilis*,) with specimens of the five commercial grades (1, froth of blood; 2, flower of blood; 3, 4, 5, blood of first, second, and third qualities) of the white variety, and of the round beads, negligée beads, bracelets, pins, coronets, armlets, and earrings, &c.

White coral, *Oculina*, sp., used by jewelers.

Madrepores and other showy corals, used for ornamental purposes.

Horny axis of black flexible coral, (*Plexaura crassa*,) used for canes and whips in the Bermudas.

17. CORAL—Continued.

a. Coral as a material:

Axis of fan coral, (*Rhipidogorgia*,) used for skimmers and strainers in the Bermudas.

Coral, used for building purposes.

Coral rock of recent formation, (Coquina,) used in Florida in manufacture of ornamental vases and carvings.

Calcined coral, used for dentifrices, as an antacid, &c.

Imitations of red coral in celluloid, rubber, and other substances.

18. INFUSORIAL EARTHS.

a. Polishing powders, (used for polishing metals, cabinet-ware, and stone:)

Specimens of polishing-slate, tripoli, and other foreign polishing-powder.

Specimens of American infusorial deposits.

b. Infusorial earths, employed in manufactures:

Infusorial earth, used in making window and plate glass.

Infusorial earth, used in making soluble glass.

Infusorial earth, used in making mortar.

Infusorial earth, used in making molds for metal casting.

Infusorial earth, used in making filters.

Infusorial earth, used in making dynamite.

Infusorial earth, used in making fire-proof packing.

Infusorial earth, as an absorbent for oils and liquids.

19. OTHER MATERIALS FROM INVERTEBRATES.

a. From insects:

Brazilian diamond-beetles, used in jewelry.

Wings of beetles, used in embroidery.

b. From echinoderms:

Spines of echinoids, used for slate-crayons.

** *Flexible materials.*

20. LEATHERS. (Embracing the hides in a rough state, in the various stages of dressing, and manufactured into shoe-leather, parchment, vellum, binder's leather, thongs, &c.)

a. Prepared from mammal skins:

Cat-leather.

Dog and wolf leather, used for drum-heads, &c.

20. LEATHERS—Continued.

a. Prepared from mammal skins:

Bear-leather.

Raccoon-leather, used for gloves and upper-leathers of shoes.

Seal-leather, used for fine shoes and in the manufacture of "patent leather," and by Eskimos for numerous purposes.

Sea-lion leather, used by Eskimos to cover bidarkas and for garments and beds.

Walrus-leather, used by Eskimos for harness, tables, thongs, seal-nets and for covering polishing wheels.

Bison-leather (and buffalo-leather, buff-leather.)

Ox-leather, with specimens of sole-leather, split-leather, grain-leather, rawhide thongs, whips, leather-belts and saddles, and of calf-skins, prepared for binders' and bootmakers' use, as Russia leather and vellum, and tawed, as parchment.

Sheep-leather, with specimens of binder's leather, imitation chamois leather, wash-leather, buff-leather, roan, imitation morocco and parchment, with vellum made from skins of dead-born lambs, and manufactured gloves, &c.

Goat-leather, with specimens of shagreen-leather, morocco-leather, as used for linings, upholstery, bindings, and pocket-books, parchment, drum-heads, &c., with kid-leather, used in manufacture of shoes and gloves, underclothing, and vellum made from skin of young kids, also skin-bottles used in Asia.

Horse and ass leather, used in manufacture of shagreen, sole-leather, harness-leather, saddles, trunks, water-hose, pump-valves, military accoutrements, ladies' shoe-uppers.

(Chamois leather, (*Capellà rupicapra,*) used for polishing purposes and for straining mercury.)

(Leather of gazelle, (*Gazella dorcas,*) used in packing commercial aloes, and of African antelopes, used in packing elephants' tusks.)

Deer-leather, dressed as buff-leather, chamois-imitation leather, Indian dressed (buckskin,) and for the finer moroccos, also manufactured into gloves, gaiters, undergarments, polishers, &c.

Moose-leather in ordinary and buckskin finish.

20. LEATHERS—Continued.

a. Prepared from mammal skins:

Caribou-leather in ordinary and buckskin finish.

(Reindeer-leather.)

Elk-leather in ordinary and buckskin finish.

Antelope-leather in plain, buckskin, and oil-finish, used in manufacture of castor-gloves.

Peccary-leather as used in the manufacture of gloves.

Hog-leather used by saddlers, shoemakers, and book-binders.

Hippopotamus-leather used for buffing or polishing wheels.

Rhinoceros-hide used for shields, targets, whips, &c.

Beluga-leather dressed as kid, sole, harness, velvet, plush, boot, mail-bags, belts, and patent (varnished) leather.

Porpoise-leather.

Beaver-leather used in manufacture of saddles, shoes, gloves, and trunks.

(Nutria-leather (*Myopotamus coypus*) of South America.)

Rat-leather used for thumbs of kid gloves.

(Kangaroo-leather.)

Leather trimmings used as stuffing for balls, &c.

b. Prepared from intestines of mammals:

Parchment from viscera of seals, used by Eskimo for clothing, bags, and blankets.

Leather from pharynx of seal and walrus used by Eskimo for boot-soles.

Parchment from viscera of bears used in Kamtchatka for masks and window-panes.

Viscera of ox used in manufacture of gold-beaters' skin.

Bladders of animals used for pouches, parchment, bottle and jar covers, and by Eskimo for oil-bottles.

Viscera of sheep used in manufacture of "cat-gut," with specimens of whip-cord, hatters' cord, for bowstrings, clock-makers' cord, filandre, guitar, violin, and harp strings, angling-lines, &c.

Viscera of hog used as envelopes for minced meat, sausages, &c.

Sinews of sheep, deer, goat, buffalo, seal, walrus, and other animals used in manufactures of threads, lines, nets, and snow-shoes, in strengthening bows, &c., the Babiche of the Eskimos of the northwest coast.

20. LEATHERS—Continued.

c. Prepared from bird-skins: (Eskimos.)

Eider-leather.

Auk-leather.

(Ostrich-leather used by Arabians.)

d. Prepared from reptile skins:

Alligator-leather.

Rattlesnake-leather.

Other snake-leather.

e. Prepared from fish-skins:

Leather prepared from scaled fish by Indians.

Eel-leather, (pigtails, queues, flail-thongs.)

Shark-leather, (shagreen used for coverings and by the Alaska Indians for boot-soles.)

Sturgeon-leather.

(Skins of *Diodon* used in making helmets.)

Stomach membranes of halibut used in Greenland for window-transparencies.

f. Leather waste:

Paper manufactured from waste.

Glue manufactured from waste, (see under 24.)

Prussian blue made from leather waste, (see under 30.)

21. HAIR AND WOOL.

a. Hair used in weaving and felting, (see under 8.)

b. Hair used for wigs and ornament:

Human hair as an article of commerce, with specimens of switches and wigs, and also of the trade imitations of hair in jute, horse-hair, &c.

Goats' wool as employed in manufacture of wigs and perukes.

Horse-hair employed for military accoutrements and for standards, (Turkey.)

Human scalp-locks as Indian trophies.

Scalps of animals as trophies.

c. Hair and bristles used for brushes, (embracing the commercial hair and bristles, assorted and unassorted, and specimens of the manufactured articles:)

Hair of skunk used for fine brushes.

Hair of bear used for varnishing-brushes.

21. HAIR AND WOOL—Continued.

c. Hair and bristles used for brushes:

Hair of American badger used for fine shaving, graining, gilding, and dust brushes.

(Hair of European badger used for coarse brushes.)

Hair of dog used for coarse pencil-brushes.

Hair of squirrel, marten, sable, kolinsky, and weasel, especially the tails, used in making fine artists' pencils.

(Hair of camel used for pencils.)

Bristles of hog and peccary used in making coarse brushes for varnishing, scrubbing, &c.

Tails of horses, buffaloes, &c., used for fly-brushes.

(Tails of yak used for fly-brushes.)

(Tails of elephants used for brushes and standards.)

Sheep's wool (on skin) used for black-board rubbers.

Hair of deer and antelope (on skin) used by Indians for hair-brushes.

Ox-hair from the inside of cows' ears used for striping and lettering brushes.

d. Hair used in other manufactures:

Bristles used in shoemakers' waxed ends.

Bristles used in anatomical instruments.

Hair and bristles used in artificial flies. (See under B, 45.)

Hair of cattle used in strengthening mortar and plaster.

e. Hair used for stuffing:

Horse-hair, straight and curled, used for mattresses and cushions.

Refuse hair of beaver and musquash, cut from felting-hair, used for cushions.

(Down of rabbits used for cushions.)

f. Wool used as a medium for pigments:

Wool flocking used in the manufacture of wall-paper, colored felts, and rubber-cloth.

g. Chemical products:

Refuse human and other hair used in manufacture of prussiate of potash, with specimens of manufactured product.

22. QUILLS.

a. Quills of mammals:

Quills of American hedge-hog used by Indians in embroidering.

22. QUILLS—Continued.

a. Quills of mammals:

(Quills of porcupine used for pen-holders, floats for fishing, eyelet-punches, &c.)

(Quills of European hedge-hog, on skin, used as a muzzle for weaning calves.)

b. Quills of birds:

Quills of swan and turkey for engrossing-pens.

Quills of goose and eagle for writing-pens.

Quills of crow and duck for fine pens.

Quills used in making toothpicks, fishing-floats, color-bottles, pencil-handles, needle-holders, &c.

23. FEATHERS.

a. Feathers used for clothing. (See under Furs, D 6.)

b. Feathers used for implements, (including manufactured articles:)

Feathers of hawks used as fans and screens.

Feathers of fowl, turkey, grouse, and peacock used for brushes, fans, and screens.

Feathers of ibis, spoonbill, egret, and bittern used for fans and screens.

Feathers of flamingoes, swans, geese, and ducks used for fans and screens.

c. Feathers used for plumes and ornament, (including plumes, head-dresses, cockades, hat and dress trimming, &c.:)

Feathers and wings of small perchers used in millinery and in manufacture of feather flowers.

Feathers of trogons and birds of paradise used as plumes and for feather flowers.

Feathers of humming-birds, scalps, and throats used in ornamental work.

Feathers of kingfishers used in plumagery.

(Feathers of parrots used in making feather flowers.)

Eagle and hawk feathers used for plumes.

Feathers of pigeons used for ornamental work.

Feathers and wings of cock used as plumes, trimmings, &c., natural and dyed.

Breast feathers of grouse, pheasants, and turkeys used as roll-plumes in hats.

23. FEATHERS—Continued.

c. Feathers used for plumes, &c.:

Feathers of ibises, spoonbills, flamingoes, herons, egrets, and bitterns used for plumes and ornamental work.

(Feathers of adjutant, (*Lepoptilus argala,*) and marabou, (*Lepoptilus marabou,*) used for plumes and trimmings.)

Feathers of flamingoes, swans, geese, and ducks used in ornamental work for roll-plumes, and swans' down for trimmings. (See under 6.)

Breast-feathers of gulls, terns, and tropic birds used as roll-plumes.

(Feathers of African ostrich used for plumes and trimmings, with specimens of undressed, scoured, bleached, scraped, and dyed grades.)

Feathers of American ostrich.

Specimens of composite feather flowers.

Specimens of plumagery work on metal.

Specimens of birds mounted for use in millinery.

d. Feathers used in other manufactures:

Feathered arrow-shafts. (See under B, 18.)

Feathers used in making artificial flies. (See under B, 45.)

Feathers used in manufacture of textile fabrics. (See under D, II, C.)

e. Down of birds:

Down of eider-duck used in bed-stuffing, with specimens of the balls in which it is packed for transportation.

Down of other ducks.

Down of geese and swans used as stuffing for beds, and as electrical non-conductor in manufacture of philosophical instruments.

24. GELATINE AND ISINGLASS.

a. Gelatine:

Gelatine made from leather-shavings, bones, hoofs, and horns of bison, cattle, sheep, and other domestic animals, used in manufacture of glue, size, court-plaster, *papier glacé* for tracing, imitation glass, artificial flowers, and ornamental work, wrappings for confections, table-jelly, (see under D. 1,) &c.

Size and gelatine from fine ivory chips.

24. GELATINE AND ISINGLASS—Continued.

a. Gelatine:

Bone-glue, (Osteocolla.)

(Glue made in India from skin of the ass, (Hippocolla.))

b. Isinglass:

Isinglass, (Ichthyocolla,) made from air-bladders and skins of fishes and used in the manufacture of fine glues and sizes, adhesive and court plasters, diamond cement, imitation glass, and table-jelly and confectionery, (see under D. 1, D,) in refining wines and liquors, in adulterating milk, in fixing the luster of artificial pearls, and in lustering silk ribbons, (embracing the dried bladders and the manufactured products,) in their grades of "lyre," "heart-shaped," "leaf," and "book" isinglass.

Isinglass from sounds of cod and hake.

Isinglass from the squeteague family, (*Sciœnidae,*) principally used by confectioners.

Isinglass from cat-fish family, (*Siluridae.*)

Isinglass from carp family, (*Cyprinidae.*)

Isinglass from sturgeons in all its grades and commercial forms.

Isinglass prepared from fish-skins.

25. FLEXIBLE MATERIALS DERIVED FROM INVERTEBRATES.

a. Insect productions:

Silk-worm "gut" used in making leaders for fish-lines.

(Nest of Cayenne-ant, (*Formica bispinosa,*) used as a mechanical styptic.)

Spiders' web used as a mechanical styptic and for the cross-lines in optical instruments, (see, also, under D, 8.)

Papier-maché of hornets' nests used for gun-wadding.

b. Mollusk productions:

Byssus of mollusks, (see under D, 8.)

26. SPONGES.

a. Specimens of American commercial sponges, with the different grades, and bleached sponges:

(Specimens of Mediterranean sponges.)

Surgical apparatus, probangs, aurilaves, "sponge-tents," and other instruments manufactured.

Spongeo-piline used as a substitute for poultices.

Sponges used in stuffing mattresses and cushions.

27. OILS AND FATS.

a. Mammal oils:

Bear-oil and bear-fat used as a cosmetic and in the manufacture of pomatums.

Dog-oil used in the manufacture of kid-gloves.

Seal-oil, in its various grades, used for lubricating.

Sea-elephant oil.

Sea-lion oil.

Manatee-oil.

Dugong-oil.

Oil and fat from domestic animals, tallow, suet, lard, lard-oil used in lamps, for lubricating, and neats-foot oil used in dressing leather, also manufactured into various substances, (see D, 30,) and tallow candles and night-lights.

Oil from body of whales, grampuses, and porpoises used in the arts, for lubricating, painting, &c.

Black-fish and porpoise-jaw oil used in lubricating fine machinery, watches, clocks, and guns, with specimens of blubber.

Grampus-oil used for lubricating fine machinery.

Sperm-oil used in lamps, for lubricating, as an emollient in medicine, for lip-salves, and in the manufacture of spermaceti.

Manufactured glycerines, used as a preservative and antiseptic, as a cosmetic, as an emollient, as a substitute for cod-liver oil, in the manufacture of perfumes and hair-dressings, in photography, in the manufacture of nitro-glycerine, dynamite, dualine, lithofracteur, coloniamite, and other explosives, soap, &c.

Manufactured stearines, with candles and other manufactured articles.

Soaps manufactured from mammal-oil, soda-soaps, (hard, toilet, and resin soaps,) potash-soaps, (washing, shaving, and soft soaps,) diachylon plaster, &c.

Spermaceti, with specimens of candles.

Butter made from milk of cows, goats, and horses.

Oleomargarines, with specimens of imitation butter.

Brains of buffalo used in tanning by Indians.

b. Bird-oils:

(Oils of petrels and other sea-birds used by Eskimos and in the Azores for lamp-oil.)

27. OILS AND FATS—Continued.

b. Bird-oils:

Goose-oil used by watch-makers, and as an emollient.

(Oil of guacharo, (*Steatornis caripensis*,) used in South America as food.)

(Ostrich used for food, and by the Arabs in medicine, and emu-oil used in Australia in medicine.)

(Oil of penguin, (*Diomedea chilensis*,) of Falkland Islands, sold in London for currying leather.)

(Peacock's fat and oil.)

(Oil of mutton-bird, (*Procellaria obscura*,) of Bass's Straits, used for lamp-oil illuminating.)

(Oil of frigate-bird, (*Tachypetes aquila*,) sometimes used in medicine.)

Oil of pigeon, (*Ectopistes migratorius*,) used as food by Indians and frontiersmen.

(Fulmar-oil from island of Saint Kilda.)

c. Reptile oils:

Alligator-oil manufactured in Florida.

(Alligator-oil used by South American Indians, mixed with chica pigment for painting their bodies.)

Turtle-oil made from turtle-eggs, used in dressing leather and in manufacture of soap.

Rattlesnake and other snake oils.

d. Fish-oils:

Sun-fish oil used by fishermen for cure of rheumatism.

Cod-oil, also cod-liver oil used in medicine, as a food and emollient, and in lubricating.

Hake and haddock liver oil used in adulterating cod-liver oil.

(Pollock-oil used by Shetlanders for illumination.)

Menhaden-oil used in currying leather, in rope making, for lubricating, for adulterating linseed-oil, as a paint-oil, and exported to Europe for use in the manufacture of soap and for smearing sheep.

Herring-oil.[1]

White-fish oil.[1]

Sturgeon-oil.[1]

([1] NOTE.—These oils, with other oils made from fishes, and a large part of the seal and "black" whale oil, are known indiscriminately as fish-oil and used chiefly for the purposes enumerated under the head of menhaden-oil.)

27. OILS AND FATS—Continued.

d. Fish-oils:

Oulachan oil used by Indians of Northwest coast for food and illumination.

Shark and skate liver oil, including the "Rouen oil," made on the coast of Normandy from the livers of *Raia aquila*, *R. pastinaca*, and *R. batis*, used like cod-liver oil.

Cramp-fish oil used by fishermen for cure of rheumatism.

Soaps made from fish-oil.

28. PERFUMES.

a. Mammal perfumes:

(Civet of the civet-cat (*Viverra civetta*) of Africa.)

(Civet of the rasse (*Viverra rasse*) of Java.)

(Zibeth civet of the Zibeth (*Viverra zibetha*) of Indian Archipelago.)

(Musk from musk-deer, (*Tragulus*, sp. var.,) in its various grades, of Tonquin or Thibet, and Kabardin, Russian, or Siberian musk.)

Musk of musk-ox.

Musk of the musquash.

Castoreum of the beaver, including the various commercial grades, the Canadian, Hudson's Bay, and Russian castoreum, and specimens of castorine.

(Hyraceum of the daman, (*Hyrax capensis*.))

Ambergris of sperm-whale, with specimens of ambreine.

b. Reptile perfumes:

Musk of alligator.

Oil of hawksbill and loggerhead turtles, used in perfumery.

29. COLORING MATERIALS.

a. Derived from mammals:

Bone-black.

Ivory-black, (*noire d'ivoire*,) used in fine painting, and in the manufacture of bank-note ink.

Prussiates, prussian-blue, ferrocyanide of potassium, made from hoofs and refuse human and other hair.

Gall of animals used in dyeing.

Dung of animals used in calico-printing.

Hæmatin made from blood, and used in turkey-red dye-works, and for the red liquor of printers.

Wool-flocking. (See under D, 21.)

29. COLORING MATERIALS—Continued.

b. Derived from birds:

Shell of eggs, used for white pigment.

Series of murexides, or purpurate of ammonia dyes, made from guano.

c. Derived from fishes:

(*Essence d'Orient*, or fish-scale pearl, used as a pigment.)

(Gall of carp, used in Turkey as a green paint and in staining paper.)

d. Derived from insects:

(Cochineal dye, from *Coccus cacti* of Mexico, used in manufacture of rouge, of carmine, and lake pigments, and in coloring tinctures.)

Canadian cochineal.

(Kermes(and other cochineals of commerce, *Coccus ilicis.*)

(Lac dye and lac lake, from *Coccus lacca*, *C. polonicus*, *C. uva ursi*, and *Ophis fabæ.*)

Dye prepared from bed-bug, (*Cimex lectularius.*)

(Dye prepared from *Trombidium*, in Guinea and Surinam.)

Nut-galls produced by insects, and used in tanning, for black dyes, for woolen cloth, silk, and calico, and in manufacture of ink and gallic and pyrogallic acid, employed in photography.

e. Derived from mollusks:

(*Sepia* from *Sepia officinalis.*)

Purple dyes from gastropods, *Murex*, *Purpura*, &c.

Purple dyes from nudibranch mollusks.

30. CHEMICAL PRODUCTS AND AGENTS EMPLOYED IN ARTS AND MEDICINE.

a. Derived from mammals:

Secretion of skunk.

Album græcum of dogs, used as a depilatory in tanning hides.

Albumen of blood, employed in sugar-refineries, in certain cements and pigments, and as antidote and emollient.

Dung, used in calico-printing.

Gall of animals, used in mixing colors, in fixing the lines of crayon and pencil drawings, in preparing the surface of ivory for painting, in removing grease, and in medicine.

Pepsine and pancreatine, prepared from stomachs of hogs and calves.

30. CHEMICAL PRODUCTS, &c.—Continued.

a. Derived from mammals:

(Koumiss, a fermented liquor, prepared from mare's and cow's milk, and employed in medicine.)

Phosphorus, prepared from bones, with specimens of matches, vermin poisons, and other products.

Vaccine lymph, derived from cows.

Ammonia, prepared from bones and horn.

Sal ammoniac, prepared from bones and dung.

Prussiates, prepared from hoof, horn, and leather waste, dried blood, hair, and wool, with specimens of blue cyanide of potassium.

Lime from bones and bone phosphates. See also under 32.

Punk and tinder, made from droppings of camel and bison.

Animal charcoal, used as a decolorizer.

b. Derived from birds:

Albumen of eggs, used in photography, in clarifying liquors, by physicians as emollients and antidotes, and by apothecaries in suspending oils and other liquids in water.

Egg-shells, employed as an antacid.

c. Derived from reptiles:

Crotalin of rattlesnake and copperhead.

(*Scincus officinalis* of Egypt, used by European practitioners as sudorific and stimulant.)

d. Derived from fishes:

Propylamine, made from fish-brine.

(Intestines of grayling, used by Laplanders as a substitute for rennet.)

Skins of eels, used by negroes for rheumatism.

e. Derived from insects:

Vesicatory preparations from American beetles, *Cantharis cinerea* and *C. vittata.*

(Vesicatory preparations derived from foreign beetles, cantharides or Spanish flies, (*Cantharis vesicatoria*,) and other species, and substitutes *Mylabris cichorii*, *Cercoma Schoefferi*, *Meloe*, sp. var., &c.)

Vesicatory preparations from American spiders, such as *Tegenaria medicinalis.*

Gall-nuts, used in medicine. (See under 29.)

30. CHEMICAL PRODUCTS, &c.—Continued.

e. Derived from insects:

Coccinella, used as a remedy for toothache.

(Trehala, made from nests of beetles, (*Larinas nidificans*,) of East Indies, and used as a substitute for tapioca.)

Formic acid.

Carbazotic acid and its derivatives, made from sewing silk scraps, and used as a substitute for quinine.

Beeswax, used in manufacture of candles, cerates, plasters, and artificial flowers, in modeling and casting, and in medicine.

Honey, used as a preservative, a food, and in medicine as an aperient and demulcent.

(Wax, used in Chinese pharmacy, secreted by the *Coccus pehlah*.)

(Manna, produced by punctures of *Coccus manniparus*.

a'. Manna from the *Tamarix mannifera*, used as food, and in medicine as a purgative.

b. Cedar manna of Mount Lebanon, from *Pinus cedrus*.

c. Arabian manna, from *Hedysarum alliagi*.)

(Eye-powder, made by Chinese from the Telini fly, (*Mylabris cichorii*,) of India.)

f. Derived from crustacea:

Salve-bug of fishermen of Banks, (*Caligus curtus*,) parasite on cod-fish.

Crabs' eyes, or concretions from stomach of astacus, used as an antacid.

g. Derived from worms:

American leech, (*Macrobdella decora*,) used in surgery.

(European leech, (*Hirudo medicinalis*,) introduced into America.)

(African leech, (*Hirudo trochina*,) introduced.)

Leeches used as barometers.

h. Derived from mollusks:

(Cuttle-fish bone of *Sepia officinalis*.) (See under D, III, H.)

Calcined shells, used for building-lime, and in manufacture of dentifrices and enamel. (See under III, H.)

i. Derived from radiates:

a. Limes, derived from calcining coral and coral rock.

30. CHEMICAL PRODUCTS, &c—Continued.

k. Derived from protozoans:

Burnt sponge, formerly used in medicine.

Infusorial earth, and its applications. (See above, under **K.**)

31. FERTILIZERS.

a. Natural guanos:

Bat guano from caves.

Bird guano from oceanic islands.

b. Artificial guanos:

Menhaden guano.

Herring guano.

White-fish guano.

Other fish guano.

c. Artificial fertilizers:

Bone-dust ground for use.

Bone phosphates.

Dried meat from refuse of slaughter-houses.

Poudrettes.

Other animal fertilizers.

32. LIMES. (See under 30.)

33. OTHER MATERIALS NOT MENTIONED.

SECTION E.

PROTECTION AND CULTURE.

I. INVESTIGATION.

1. METHODS OF THE UNITED STATES FISH COMMISSION.

a. Methods of work:

Apparatus for collecting specimens, (see under B.)

Apparatus for physical research.

Appliances for working up results.

(This should include a model of coast laboratory with all its fittings.)

b. Results of work:

Publications of the commission.

Collections, (see under A, V to VIII.)

Photographs, &c.

II. PROTECTION

2. PRESERVATION OF GAME, FISH, &c.:

* *From man.*

a. Game laws.

** *From artificial obstructions.*

b. Fish-ways:[1]

Gap fish-ways.

Trench, ditch, or "Cape Cod" fish-ways.

Oblique grove fish-ways:

Single groove.

Brewer's.

Mather's.

Step fish-ways:

Box or pool fish-ways:

Overflowing, (old style.)

With passage-way cut down to the floor, (Smith's.)

With passage-way submerged, (Cail's.)

With contracting galleries, (Pike's.)

With transverse-sloping floors, (Steck's.)

---

[1] Classification proposed by C. G. Atkins.

2. PRESERVATION OF GAME, FISH, &c.—Continued.

   b. Fish-ways:

   Steps contrived by arrangement of rocks and bowlders.

   Inclined plane without steps:

   Plain, (Pennsylvania.)

   With partitions at right angles:

   "Rectangular compartment."

   Brackett's.

   With oblique partitions:

   Foster's.

   Swazey's.

   *** *From natural enemies.*

   c. Apparatus for destroying injurious species:

   Oyster-bed tangles, (see under B, 12.)

3. CARE OF ANIMALS IN CAPTIVITY.

   a. Tethers and hopples.

   b. Cages and pens:

   Kennels for dogs, &c.

   Cages for animals.

   Cages for birds.

   Cages for insects.

   (West India fire-fly trap.)

   c. Fish-cars and other floating-cages for aquatic animals.

   d. Aquaria:

   Globes.

   Aquaria.

   e. Hives and other cages for insects.

   f. Live-boxes, troughs, &c., for microscopists' use.

   g. Fish-ponds, fish-farms, (models.)

4. ENEMIES OF USEFUL ANIMALS.

   a. Intestinal worms and other internal parasites.

   b. Fish-lice, barnacles, and other external parasites.

   c. Predatory animals not elsewhere exhibited.

III. PROPAGATION.

5. PROPAGATION OF MAMMALS.

   a. Methods of mink culture.

   b. Methods of culture of domesticated animals.

6. PROPAGATION OF BIRDS.

a. Methods of ostrich culture.

b. Methods of culture of domesticated birds, fowls, &c.

7. PROPAGATION OF REPTILES.

a. Methods of terrapin culture.

8. PROPAGATION OF AMPHIBIANS.

a. Methods of frog culture.

9. PROPAGATION AND CULTURE OF FISHES.[1]

a. Accessories of obtaining and impregnating ova:

Pans, pails, &c.

Strait-jackets used in spawning salmon.

Spawning-race, (Ainsworth.)

Roller-spawning screen, (Collins.)

Spawning-vat, (Bond.)

b. Hatching-apparatus:

Troughs:

Plain.

Gravel-bottomed.

With sieve-bottom trays:

Brackett's.

Williamson's.

Clark's.

Vats or cases:

Holton's.

Roth's.

Glass-grilled boxes, (Coste's.)

Jars and tin-vessels:

Bell and Mather's.

M. A. Green's.

Ferguson's.

Chase's.

Hatching-boxes, (floating:)

Seth Green's shad-box.

Brackett's shad-box.

Brackett's shad-box, (No. 2.)

Bryant's shad-box.

Stilwell & Atkins's shad-box.

Bannister's shad-box.

---

[1] Classification proposed by J. W. Milner.

9. PROPAGATION AND CULTURE OF FISHES—Continued.

b. Hatching-apparatus:

- Hatching-boxes, (floating:)
  - Adhesive eggs apparatus:
    - Vertical wire-cloth trays.
    - Hatching-basket.
- Brook shanty, (Furman's.)
- (Bay or cove barriers, Professor Rasch's.)
- Accessories:
  - Tanks.
  - Nests.
  - Trays.
  - Grilles.
  - Gravel-filters.
  - Flannel screens.
  - Shallow troughs or tables (for picking eggs.)
  - Egg-nippers.
  - Cribbles.
  - Pipettes.
  - Skimmer-nets.
  - Feathering quills and brushes.
  - Rose-nozzles, (for washing eggs.)
  - Syringes, bulb, &c.
  - Shallow pans.
  - Aerating-pipe.

c. Transporting apparatus:

- Apparatus for transporting eggs:
  - Cans.
  - Case of cups, (Wilmot's.)
  - Case of cups, (Clark's.)
  - Case of trays, (Clark's.)
  - Moss-crates, (Stone's.)
- Apparatus for transporting fish:
  - Barrels.
  - Cans, plain.
  - Cans with aerating accessories:
    - Slack's.
    - Clark's.
    - Creveling.
    - M. A. Green's.

9. PROPAGATION AND CULTURE OF FISHES—Continued.

c. Transporting apparatus:

Apparatus for transporting fish:

Tanks with aerating accessories:

Tanks, with attachment of band-wheel to car-axle, (Stone's.)

(Tanks, with Freiburg aerating apparatus.)

Aquarium-car, (Stone's.)

Live-box, (Atkins's.)

Accessories:

Air force-pumps.

Siphons.

Siphon-tubes.

Bellows.

Roses, aerating.

10. PROPAGATION OF INSECTS.

a. Propagation of silk-worm:

Specimens of plants used for food.

Model of house and its appliances.

b. Propagation of cochineal insect.

c. Propagation of bees:

For hives, (see under E. 3.)

11. PROPAGATION OF WORMS.

a. Propagation of leeches.

12. PROPAGATION OF MOLLUSKS.

a. Methods of oyster culture:

Stools for receiving spat, natural and artificial.

Other apparatus.

13. PROPAGATION OF CORALS.

14. PROPAGATION OF SPONGES.

INDEX.

www.ingramcontent.com/pod-product-compliance
Lightning Source LLC
LaVergne TN
LVHW021417110826
845150LV00007B/1961

* 9 7 8 1 4 2 5 5 0 9 9 1 0 *